Origins of
Suffolk Place-Names

Origins of Suffolk Place-Names

Compiled by Mel Birch
including work on the subject by the late Claude Morley

a

mini-series publication

ISBN 0 948134 63 1

Cover Illustration:
Laxfield from a watercolour original by Hamilton Murray-Hill

CASTELL PUBLISHING © 2003
Castell Publishing Mendlesham Suffolk IP14 5RY

This book is sold subject to the condition that it shall not, by way of trade or otherwise, be lent, hired out, or otherwise circulated, without the publisher's prior consent, in any form of binding or cover other than that in which it is published, and without a similar condition being imposed on the subsequent purchaser.

Introduction

The Hundreds of Suffolk

With unthinkable arrogance, the government of 1974 swept away the prehistoric hundreds and redrew the map of the English counties to 'tidy up' boundaries but also, no doubt, for their own short-term electoral advantage and by no consent of its people. It is therefore comforting to know that, unlike governments, these ancient territorial boundaries will forever live down History, for nothing over a thousand years old can be entirely eradicated.

Until recent times it was accepted that since the Anglo-Saxons distributed their land into hundreds, or their equivilent, before they left the Continent, it was logical to believe that they transplanted a similar system on this country when it had been sufficiently subjugated. But even as ealy as 1895 Raven had pointed out that 'although our divisions into hundreds are said to date from Alfred the Great, perhaps some more settled districts got into shape earlier than others less favoured.'

Alfred the Great was never to wield more than the most nominal control over East Anglia and his biographer at the time, Asser, never referred to hundreds as an innovation. On the contrary, hundreds are actually recorded among the Continental Germans by Tacitus circa AD. 80 hence the probability that these English divisions came into being as soon as enough settlers were congregated to form so large an assembly. The defensive military organisation of the Danish intruders, the Wapentake, was never sufficiently potent in East Anglia to supersede the hundreds. A much more likely origin was suggested by H. C. Coote who wrote in 1876 that they are 'in my opinion, to be identified with (Roman) milites stationarii; these were police, were decimal in their organisation, and were posted in stations which received their names through the same system, from the cohortal divisions of the force, viz. centuriarie and decanica. Further, these stations were placed within the territory of each city; and the more or less large vicus, which formed the quarters, was the centre of each police district. On the other hand, the Anglo-Saxon hundred and tithing were police districts, taking their names from the vills and villages which formed their centre. They were divisions of the shire, which itself was conterminous with and no other than the territorium of a Roman city. Further the names hundred and tithing were necessarily a reminiscence of a numeral system which, and which only, could have given them such a designation. Gothic peoples reckoned by the long hundred, and its innate practice by our Saxons throughout their occupation has endured to our own days. If one allows such perpetuation from the Romans, the titles of Suffolk hundreds began during Pagan days, if not from actual Latin.'

It is hardly likely that the local population still felt oppressed after four-hundred years of Roman rule. A great majority now had Roman blood flowing through their veins and all had known Roman 'rule' as the norm. By this time it was unlikely they felt governed by foreigners at all, and it is known that many English had positions of responsibility in the Roman hierarchy. It is therefore false to assume, as many have, that when the militia departed, people returned to their pre-Roman barbarism (if it were ever so). The Romans may not have introduced the hundreds but they must have policed the country by dividing it into managable sections - probably of hundred size. It is possible these divisions were later taken up by the Saxons, likely using the same (or similar) boundaries but renaming each from a particular landmark where the Hundred Courts would meet - a river, ford, tumulus, etc. The gap between the loss of Roman control and a reasonably sized occupation by the Saxons is thought to be around 200 years - the divisions would hardly have been forgotten in that time and much more likely have remained on-going throughout this period. The Dark Ages are often taken as a time of anarchy, but the lack of records are more likely due to problems caused by Saxon and Danish insertion than any increase in lawlessness among the local population who no doubt carried on their day to day life much as they had done before. The Danes may have been intent on pillage but the Saxons ultimate aim was not conflict but land to farm and settle.

Babergh Hundred : The first of our twenty-one Hundreds lies in the south-west of the county with the river Stour separating its western and southern boundaries from Essex. Babergh doubtless was once the name of a township, though not so in 1086 when first we obtain a clear view of Hundreds' exact distribution and extent. Now it is the name of a heath in Great Waldingfield, with a hall with ancient earthworks on its east, which may be considered to mark the site of the Hundred Court. The official form Babergh, shown to be 'Bada's hill' by Skeat, appears in the Hundred Rolls of 1275, along with the better spelled *Badberewe*, which becomes Baddebury Hundred in Inq. post mortem. Variability in Domesday spellings is, not unreasonably, rife, and under Eleigh (alone) are entered the Hundred of *Babenberga, B'hinberga, Banberga,* and *Babergh*, all incorrect, for Kemble's Index shows the true AS. *Baden beorh* 'the hill mound or tumulus of Bada'. But the one of 1304 surely leaves but small doubt that it is the god *Balder* or *Baldaeg* that is here commemorated.

Blackbourn Hundred : Blackbourn lies in the N of the county to the E of Lackford Hundred and takes its name from the river Blackbourne, a section of the river Thet. The Blackbourn derives its name from the dark colour of its peaty water. The Hundred is also watered by the Lark and the Little Ouse rivers. The smaller streams all flow into the Ouse. The site of the Hundred Court was thought to have been in Stanton village. In Domesday Survey this division of the county appeared as the double Hundred of Blakeburn and Bradmere, names which are perhaps still perpetuated in the modern Blackburn Farm in Stanton and Broadmere in Troston. The division passed through Bardwell village.

Blything Hundred : Extending about twenty miles along the coast from Aldeburgh to Southwold (East Bavents), Blything Hundred is bounded on the north by Wangford and Mutford Hundreds and on the south by Plomesgate. The river Blyth was of paramount importance when this Hundred was created. Blythburgh with its erstwhile broad estuary to the sea is just the place upon which early descents might be made by the first Angles who would, consequently, be termed Blithingas or "dwellers beside the Blyth", a pleasant stream, an affluent of which runs through Holton Vale. No doubt the Hundred Court met at Blythburgh. The northern part of this Hundred later composed that of Mutford.

Bosmere and Claydon Hundred : This Hundred extends from the Borough of Ipswich to Badley and Creeting on the border with Stow Hundred, and from Offton and Willisham on the borders of the Cosford Hundred to Helmingham. The river Gipping flows through it and at Bosmere expands into an ancient lake or mere from where the Hundred derives its name. Just E of the river

Gipping, the rising ground above the mere (locally pronounced 'Bors-mur') doubtless served the Hundred Courts. Claydon Hundred probably derived from its assemblies being at one time held somewhere by the river Gipping in Claydon township; likely where that river is now bridged to carry the road to Blakenham.

Carlford Hundred : Carlford, lying to the south of Colneis Hundred, is bounded on the west by the river Deben and Wilford Hundred, on the north by Loes and Thredling Hundred, and on the east by Bosmere and Claydon Hundreds, and by Ipswich. The river Finn rises in Tuddenham St. Martin, flowing through Witnesham, Bealings and Playford to Martlesham where it joins the river Deben. The best form of this Hundred name is the Carlesford of 1275, for the prefix is the same as that of our two townships of Carlton, meaning 'the ford of the churl or peasant'. Or we may take the Domesday Carleford to represent the AS genitive plural *carla ford*. The interesting point is that the word *carl*, for 'a rustic', is not truly Saxon but is borrowed from the Norse, with identical meaning, *karl*: but it is safer to suppose it was introduced into our language before Saxons came to Britain than by the Norse invasions of the C9. The exact location for the ford of the Hundred is not known but some point across the river Lark seemed probable. Perhaps where Mill Lane now crosses by Carlford Bridge, or more likely the present river crossing south-west of Hasketon Church.

Colneis Hundred : Colneis embraces the land between the rivers Orwell (on the south-west) and Deben (on the north-east), which is exactly tongue-shaped and was far more pointed at Walton Castle in Saxon times when mouths of these two rivers were broader than today. Hence we may believe the constituents are AS. *ness* added to some lingering Latin *collum* or *colus* about the Roman castrum that defended the headland when the sea flowed over Langer Common. The Hundred is bordered on the inland side by Carlford Hundred and Ipswich. No site of the Hundred Court seems obvious though a Calneys Manor in Falkenham and Walton is named in 1550. George Arnott postulates (1947) an eroded Colness township.

Cosford Hundred : Though many authorities give the meaning of Kersey as 'Island where cress grows', Claude Morley feels it means 'Assembly Island', the meeting place of Cosford Hundred. The village stands in an elevated position running down to the valley of the river Brett. The river now takes its name from Brettenham but was originally called Cors giving rise to Cosford Hundred and the lost Domesday township of Corsfield which lay in Babergh Hundred. When we remember that the *C* is hard and the Hundred is plainly shown in Domesday Book, the Hundred Rolls of 1275, and in IPM to have been formerly Corsford, no discrepancy but the vowel sound remains in the prefix; one was the 'island in the Cors' the other the 'ford of the Cors'. Cosford Hall in Hadleigh may well be the site of a later Hundred Court since it immediately flanks the stream, but an earlier and original Cosford Hall is shown on early OS maps as that island in the Brett which gave name to both Kersey and the hundred. Here, too, Kirby names Cosford Bridge, with a third Cosford Hall in Whatfield.

Hartismere Hundred : Hartismere Lies on the northern border divided from Norfolk by the river Waveney. In pre-Conquest times the town of Eye obviously lay amid one of the largest sheets of water in the whole county (the southern low bogland is still locally termed The More i.e. Mere) and the particular one which, almost certainly, gave name to Hartismere Hundred. The fact that the earliest form of this was *Herotesmere* seems to show *Herot* as the local lord about King Offa's time, say in AD 800.

Hoxne Hundred : This Hundred is also on the northern boundary separated from Norfolk by the Waveney. On the southern side of this Hundred is the river Alde, and the river Blyth which reaches the sea at Southwold rises near Laxfield. If, as appears likely, all Suffolk Hundreds were either perpetuated from earlier Roman Milites stationarii or established before 600 AD, we owe such later ones as this and Lothingland to mere changes of title such as is known to have happened in Mutford. Throughout all Christian Saxon times (up to c1250?), the name was **Bishops Hundred** which distinctly shows that in this case the title cannot have been given before 630 A.D. long after the Angles' advent in Suffolk. We can no more than guess from the fact that his monastery stood in Hoxne, that King Aethelbeorht gave it to the church during 790-3 A.D. The Bishop of Norwich had a country palace here.

Lackford Hundred : Occupying the north-west corner of the county, Lackford has been supposed to be a corruption of Larkford, i.e. 'the ford of the river Lark'. A portion of the Hundred is called the Fens and is part of the great Bedford Level, so called from the fourth Earl of Bedford, who commenced the work of reclaiming the fen district in 1630, a work which was completed by his son in 1653. Together they reclaiming a total of around 14,000 acres of land. Whilst Lackford parish is listed under Thingoe Hundred, the river crossing lies exactly astride the Hundred boundaries of Lackford, Thingoe and Blackbourn.

Loes Hundred : Loes is bounded by Plomesgate Hundred on the east, by Hoxne on the north and on the west and south-west by Carlford, Thredling and Wilford Hundreds. The river Deben follows a meandering course through the Hundred from Crettingham to Ufford and becomes navigable at Woodbridge. This late spelling of Loes (pronounced as in 'flows') is remarkable and Skeat considers 'it perhaps represents the AS. *Hlossan*, genitive of *Hlossa*, a personal name. If this is so, the sense refers to 'a settlement of Hlossa'. A more interesting derivation may be suggested if we accept the title 'loose,' e.g. Loose Hall in Wattisham, to represent '*Leof's*,' in which case there appears no doubt that Lose or Loes does the same. This gives us 'Loef's Hundred', and it is harder to think it an abbreviated form of Loefric's or Leofstan's, etc., than to accept the name Leof as in itself representing all that of the early owner. Prof. Earle regards Leof as a mere title or dignity, equivalent to Sire or My Lord, as applied to an Ealdorman connected with Wiltshire about the year 950 AD. Hence we at once see that this may well have been 'the Ealdorman of Eastengle's Hundred', just as Bishops' Hundred pertained to the Elmham prelates. The extent of this division has been considerably modified; in 1086 it extended from Kenton to near the Alde mouth, and included Staverton.

Mutford and Lothing Hundred : The former independent Hundreds of Mutford and Lothingland which covered the extreme north-east portion of the county have been incorporated and cover 15 miles of North Sea coast on the east. The rivers Waveney and Yare here form one of the Broads called Breydon Water and provide the northern boundary of the Hundred. On the west the Waveney bounds it from Norfolk for about nine miles. The Mutford portion lies on Wangford Hundred on the west, Blything on the south, the sea on the east and Oulton Broad and Lake Lothing on the north.

Mutford was spelt Muthford in 1263 from the AS *mutha* meaning either 'the mouth of a river' - in this case the Hundred River then a broad body of water across the much later Latimer Dam - or 'a place where two streams joined'. Beyond the erstwhile estuary we find that the formerly important road over Hulver Bridge in Mutford crosses a twin ford through the Hundred and a second small river within a few yards of each other. A point where four footpaths meet at a footbridge over the Hundred River S of Marsh Lane Farm could be the site of the Hundred Court. A field close by the ford is called 'Hanging Piece', and the road beside it, 'The Haggifer' (hanging ford?). "Villages. like kingdoms, have their period of prosperity and decay" states Suckling, "this now obscure parish was of suffcient importance in Saxon days to give name to and provide the assembly site of the Hundred."

Skeat considers Lothingland Hundred took its name from Lothing Lake. With the great story of the murder of King Edmund at Hoxne no more than some twenty miles inland, and Norse settlements thicker around Lothingland than any other part of Suffolk or Norfolk, surely the name of those murderers' father must jump to mind. Ragnar Lothbrog was slain around 855. So far as we know his 'family' who left their patronymic to our Lake, consisted of the three sons Ivar (died 873) and Ubba (died 878), both present at the Hoxne tragedy, and Halfdene (died 881). Here the suffix of Lothbrog's name has become lost. Domesday Book gives the exact area of Lothingland. It extended from Breydon Water as far south as the Hundred River between Kessingland and Benacre. The area is given in 1086 as six leagues long by two and a half and two furlongs broad. i.e 18 by 7¾ miles. The length cannot be varied to any appreciable extent since that time and, because it is only 14½ miles today, we find we have to deduct about one-fifth of the Norman measurement. So reduced, that measurement of breadth becomes about six miles. Nowadays the broadest part is eastward from St Olave's Priory - Fritton Marshes now further west were then submerged - which gives us about five miles to the coast. Now allowing the Norman coast to have run parallel with today's, (improbable though that be) we find that one whole mile has been eroded down the E side of Lothingland through the last 900 years taking with it the villages of Newton, Duneston, Rodenhall, etc.

Plomesgate Hundred :   Plomesgate follows the coast for nine miles of its eastern border. Blything Hundred lies to the north, Hoxne and Loes to the west. The river Alde is navagable as far as Snape. Domesday Book called it *Plumesgata* and the Hundred Rolls *Plumesgate* AS. *Plumes gata*. 'Plum's entry'. Here we have a masculine personal name giving genitive Plumes, elsewhere quite unrecorded. Unfortunately this is no more than the pet prefix of some name like Prun- or Brun-wine, with the -wine lost. George Arnott, who has given us the spellings Plumbyers in 1294 and obviously masc. Plumysyard in 1503, 'Plum's yard', considers it formed the eastern portion of Trimley Common.

Risbridge Hundred :   Risbridge occupies the south-west corner of the county bounded on the south by the river Stour which divides Suffolk from Essex. AS. *Hrisan bryg* - 'Hrisa's bridge'. Not solely the name of a Hundred as Skeat thought, for there is a hamlet of Barnardiston still termed Monks Risbridge, formerly extraparochial, without church, house or population. In it, however, is no stream; and no alternative Risbridge village. Nevertheless, the Stour, though not literally flowing through the hamlet itself, is close by, so it may well refer, as Skeat suggests, to a bridge which at one time crossed it.

Samford Hundred :   Samford is a triangular-shaped Hundred lying between the rivers Orwell (dividing it from Colneis Hundred) and Stour (dividing it from Essex) with the apex at Shotley to the east. Inland it borders the Cosford and Bosmere and Claydon Hundreds. This is the accepted but utterly foolish late spelling of the Domesday *Sanfort*, the *Sanforde* of 1215, and correct *Sandford* of 1275. In Suffolk it refers to a Hundred though Oxford has a township so called. The meaning is simply 'sand-ford' or a ford with sandy bottom, though across which stream within its area is now conjectural. This is utterly distinct from 'Sandesford' Manor in Waldingfield Magna (Copinger 1905) whose true name of 'Staneforde', i.e. 'Stone Ford' emerges in 1358.

Stow Hundred : The central Hundred of Suffolk. The river Gipping which rises in the hamlet of Mendlesham Green flows through the Hundred and was navagable from Stowmarket to Ipswich. Stowmarket, the *Stow* - 'meeting place (of the Hundred)' had 'Market' added by 1253. Stowmarket was earlier a royal manor called *Thorney,* a name which shows that the original settlement was on an island or peninsula in the river Gipping.

Thedwestry Hundred : A small Hundred lying in the centre of the county, Thedwestry has Bury St Edmunds on its western border. Here are the sources of several small Suffolk rivers amongst them the Rat which rises in Rattlesden and joins the Gipping at Stowmarket. The title of the Hundred should not have been encumbered with a modern terminal *y*; for the old spellings are *Thedwardestre,* and in one case *Thedwardistree*; Domesday Book has *Theodwardestreo*. The original Saxon word was *Theodweardes treo*, meaning 'Theodward's tree'. We cannot now tell just where this tree, which doubtless shaded the Hundred Court, used to stand; no such township exists, though a 'Thedwastree hill' was located between Thurston and Pakenham villages, spelled Thedwestry Hill under Thurston by William White in 1844.

Thingoe Hundred : This Hundred lies to the south of Lackford in north-west Suffolk. The river Lark rises near the southern boundary of the Hundred and flows northwards along its eastern boundary. Morley finds no historic data upon which to base Isaac Taylor's "probable" supposition (1878) that Ixworth in Thingie was the Meeting Place of the Suffolk *Thing,* i.e 'general assembly of clans'. Distinct from that of the whole county, the Thing (both AS and Old Norse, with the same meaning) of Thinghow Hundred has been placed by Gage in his *History of Thingoe Hundred* (1838) upon an artificial mound (*how*) just outside the north gate of Bury, but again with little evidence. The final *v* and *u* of Domesday Book's *Thingehov* and *Tinchou* represent *how* a small detached hill from O.Icel. *haugr* meaning 'a mound'. In both AS. and O.Norse *thing* was an assembly, a meeting for consultation or deliberation; so that every Hundred Court might be termed a 'thing' with equal propriety. Morley considered that Bradmere Hundred, now a part of Thingoe Hundred, took its name from an early broad lake, the Domesday *Bradmere* which stretched from Ixworth Priory.

Thredling Hundred : The smallest Hundred in Suffolk, Thredling contains only five parishes and was for some reason ignored in Domesday Book; the name probably appears first in IPM as *Tredelinge* with a Norman *T* for *TH* and Copinger adds *Thrydelinge* and *Thridelingge* - 'a settlement of the Thrydling family', an unknown name; or 'of the Thrythhildings' from the known AS female name *Thrythhild*. Morley had seen it ascribed to the winding course of the river Deben but could not imagine why. Its largest 'town' is the now the much reduced and isolated Debenham.

Wangford Hundred : Divided from Norfolk by the river Waveney, Wangford Hundred has on its east Mutford Hundred, with Blything Hundred on the south and Hoxne Hundred on the west. Wainford, Just outside Bungay, had a river crossing over the Waveney and was the possible meeting place for Wangford Hundred. Now merely the site of Wainford Mills it was still by way of being a township on the Waveney so late as Edward I's reign when we find a grant of land in "Wangford by Mettingham". AS. *Wain ford* 'Waggon ford' - ford wide enough and of a safe depth to take a waggon.

Wilford Hundred : Wilford is bounded on the south-west by the river Deben and the northeast by the Alde with the North Sea coast between. The location of the 1086 'Wileford' manor is lost and since 1657 the place ceased to be even a hamlet. It must have occupied an area on the E side of the river Deben opposite Melton where the present road crosses the river by Wilford Bridge (first erected 1530) and the adjacent heathland is still termed 'Wilford Square'. Skeat suggests the meaning is 'Wili's ford' or (less-likely) from the AS. *wileg* 'ford by the willow tree'. The only parish whose name relates to a ford in this Hundred is Ufford, 'the ford of Uffa (Wuffa). As this Saxon settlement possibly dates back to the C6 and relates to the important East Anglian royal family of the Wuffingus, this may perhaps be the more likely origin of the hundred place-name.

Parishes in Suffolk

'Nowadays it is fairly safe to say that the Celts and their predecessors in England were all nomads with no fixed home or, if they had one, that home was obliterated by the Roman advent. Whilst their officials did erect villas for their personal comfort, the imperial Eagles were mere conquerors in a bleak and alien land which was regarded by them solely as a mine whence to transport wealth to Italy. For such effeminate villas the succeeding Saxons had no use or respect; they were swept away. But the Saxons of whatever tribe loved his own home and stuck to it, broadly as he might roam, like a snail to its shell, thus firmly establishing a township. This he elaborated, at some indefinate period, with a bell tower to summon his family of children and farm-retainers. To this tower, after the acceptance of Christianity in the 7th-8th centuries, he added the corpus of his manorial church. This I take to be the origin of our parish.' So stated local historian Claude Morley in the middle of the last century.

Yet this is now considered a rather simplistic view of what is a much more complicated study. Trevor Rowley puts the problem more distinctly in his excellent book *'Villages in the Landscape'* when he says that 'whilst the accepted view used to be that the chronology of the English settlement could be traced through the examination of the Anglo-Saxon place-name element, recent field work has suggested that just because they had Saxon place-names, settlements had not necessarily been founded by the Saxons; therefore whilst assisting the study they must be used with caution.'

So we see that major problems confront the investigator when attempting to trace the origins of a township. Even an identifiable Saxon settlement may overlay one of Celtic or earlier origin; its ancient name having been replaced by the Saxons often using some local landscape feature. It is interesting to note that many of these natural features - rivers, woods, etc. still bear Celtic names.

Further, it has been found that evidence of previous Saxon occupation lies in close proximity to a number of isolated churches; the present settlement being some distance away and dating back no further than the Normans, say around the 12th century. The reasons for this population drift are too numerous and intricate to be encompassed within this book* but it is obvious that, in these cases, trying to identify the place-name from some incorporated or adjacent topographical feature would be a pointless exercise. Therefore until much more investigation has been carried out on the ground to locate the site of the original settlement by field-walking, many place-names must remain unidentified and undatable.

Although most of our current village names appear for the first time in Domesday Book (1086), a problem is, some do not. An additional headache is posed by the object of the Domesday Book, which was to identify taxeable estates not individual units of settlement. We cannot, therefore, be certain to what a particular location name within its pages refers; does it identify, as we are prone to suppose, our recognisable nucleated village, just a hamlet, perhaps only a single farmstead, or does it in fact apply to a group of villages.

Research become easier by the C14 when almost all the villages known to us today were in existence, together with many more that have since disappeared. We know about them from the Nomina Villarum, a document of 1316. Add to this the almost complete coverage of the Lay subsidy of 1334 and we have a clearer idea of exactly what comprised the county at that time.

An important point, and one often forgotten when trying to recreate the Suffolk of Domesday and before, is the topography existing at that time. The county would have presented a very unfamiliar appearance to us today. Firstly, and most obvious, is the coastline. Suffolk has lost thousands of acres of its vulnerable coast to the insatiable North Sea, and some of this material so savagely shorn from places such as Pakenham, Southwold, Aldeburgh and the former East Anglian capital of Dunwich, has been deposited elsewhere along the North Sea coast; at Orford, for example, it created a shingle spit which destroyed the medieval port and made the town an inland village.

Secondly, a lowering of sea levels has turned wide deep rivers into modest streams which now meander through rich, deep valleys, and returning the former island settlements of Eye and Bawdsey to mainland. Assisted in no small part by man, former fens have been reclaimed as workable agricultural land, and the great Fen Sea - shallow but still covering some fifty miles from Lakenheath to St. Ives and southwards past Freckenham, Soham and Reach to Cambridge, with Ely and a few smaller islands rising in its midst - has gone entirely to be replaced by sandy heathland. Across this sea, bordered by the (then) fishing villages of Mildenhall and Freckenham, once sailed the Norse vikings to sack Peterborough, Ely and other monasteries of the marshes during A.D. 855-80. To all this one has to add siltage from above and deduct alluvial deposits of vegetable decay from below during the past millenium, which often reveal past occupation levels lying several feet below the present soil.

There are 500 places in Suffolk by 400 different names: 97 end in *ton* (originally *tun* and sigifying early settlement, perhaps back to Roman times - Stanton (villa site) and Walton, Felixstowe (Roman fort); 81 in *ham or ingham* (a village, a village community, an estate, manor or homestead closely associated with areas of former Roman occupation - on a Roman road or in close proximity to one); 31 in *field or feld*; 26 in *ley*; 22 plus 5 Hundreds in *ford* (with higher river levels in early times, fords were often a major feature on the landscape and difficult to traverse. They were also suitable places for assembly and defensible in times of war); 17 in *ings* (place of the people or family of); 15 in *hall*; 12 in *stead* (place or site); 12 in *burh* (a defended? town or fortified place); 10 in *worth* (homestead or enclosure); 10 in *den* (a dale or valley); 9 in *ey* (an island or peninsula); 7 in *don* (a hill); 7 in *well*; 5 in *brook*; 4 in *grave* (grave or ditch); 4 in *bourn* (a stream). There are 43 unclassified. These include Groton (sandy stream?), and Hengrave (a meadow?). *By* is Danish (except with Wilby which comes from the English *beag* - 'ring of willows'), Risby, Wickham Skeith (*Skeith* is Danish for racecourse or place where horses are exercised), Thwaite (a late Danish clearing at the heart of the woods beside the Roman road). *Wick* appears to identify a Roman site close by as with Wickham Skeith (close to the Roman A140 road), Wickhambrook (near the Lydgate Roman villa), Wicken (or Wyken) Hall in Bardwell (near the Roman settlement of Ixworth). *Strad or street* refers to settlements on a Roman road, as with Stratford, Stratton Hall and Stradishall; although *street* can also refer to an outlying hamlet of a village or even a village split into two e.g. Upper Street and Lower Street. Eight end in *thorp* which is almost exclusively Norse and seems in Suffolk to be again used in the sense of a suburb (or more exactly sub-ham) of an earlier settlement; as though it were a part separated at a subsequent date. *Stoke* also appears to have the same meaning in Suffolk, a hamlet or secondary settlement. The Saxon *stoc* primarily meant a stock or log; but since it came to be used in the sense of a habitation or settlement there is little reason to doubt that the log was employed as a distinctive mark of reference, such as a tree blazed in the forest. It is a frequent place-name with all the Suffolk examples still compounds: Stoke by Clare, Stoke by Nayland, Stoke (by) Ipswich, Stoke (by the) Ash and Tostock. Despite their name, all three Suffolk *Newtons* -the one by Corton has been lost to the sea - date back at least to Saxon times. It may be assumed that they were also initially off-shoots of nearby villages which developed in their own right and built their own church. There were also manors called Newton in Akenham, Creeting and Swilland villages suggesting that, in these instances, they failed to develop and form their own identity. The original Saxon *burh* meant a fortress, a defensible stronghold, but people were sure to congregate and store their valuables under the protection of such a place, so that by sequence it came to further represent a defended town on account of its numerous inhabitants. The cause of a village's dissection into Great (Magna) and Little (Parva) was usually the presence of two or more manors within its bounds under different lords. If only one of these built a chapel, the parish all attended it, but if each erected a place of worship the homage in both a manorial and religious sense became divided also and usually has so continued to our day.

With due acknowledgement to the work of Professor Skeat, the recognised authority on Anglo-Saxon place-names, and whose derivations have been extensively used in this book when evaluating Suffolk's 500 parishes, he, along with other experts such as Ekwell, do tend to generalise. Place-names can be subjective and often open to several interpretations and Skeat rarely allows for localised historical and topographical peculiarities; he may also be less confident in his analysis outside his own period. Therefore Claude Morley's often more controversial interpretations, along with those of local historian Norman Scarfe and maritime historian George Arnott, have been favoured and used where there is dispute and uncertainty and, particularly, where it is felt a more hands-on approach to the Suffolk landscape has made their submissions more relevent to a particular site.

Place-names are listed alphabetically with their old localisied pronounciation, the hundred in which they are to be found, the roots of their origin and the interpreted meaning, the spelling of their name in Domesday Book, where possible an attempt to identify the likely site of the original settlement based on any evidence on the ground (bearing in mind the limitations placed on these interpretations discussed earlier), and general information relating to other hamlets, manors or former settlements existing or now lost in the parish.

* For a more in-depth analysis and investigation into Suffolk's original settlements and settlement movement see my forthcoming book *The Lost Villages of Suffolk*.

ORIGINS OF SUFFOLK PLACE-NAMES

ACTON *(3 m NE of Sudbury)* 'Ak-tun'
Babergh Hundred. AS. *Acan tun* - 'Farmstead or village of a man called Ac(c)a' *Achetuna* (DB). Acton Hall, N of the church, retains part of a very ancient moat in the form of a horseshoe.

AKENHAM *(3 m NW of Ipswich)* 'Ake-num'
Bosmere & Claydon Hundred. AS. *Acan ham* - 'Homestead of a man called Aca' *Acheham* (DB). Rise Hall, near the church, stands on the site of the ancient manor house.

ALDEBURGH 'Orl-brah'
Plomesgate Hundred. AS. *eald burh* - 'Old town or stronghold on the river Alde' *Aldeburc* (DB). Includes the hamlet of Slaughden 'Slorton' AS. *slog-denu* - 'Muddy valley', now virtually lost to the sea, and the lost parish of Hazlewood, pronounced 'Haz-el-wood'. Whilst no spelling occurs before the C13, it may have been a later name for the now lost Domesday township of *Nordberia* (Northbury or Northcreek), for it is said that 'in Norberia were 50 freemen belonging to Sudburne' - Sudbourne being on the opposite, southern bank of the river Alde. Perhaps *Rushmere* (in Friston) also belongs here as in 1735 Kirby speaks of "a lane called Rushmere Street over Hazelwood Common" to Aldeburgh.

ALDERTON *(8 m SE of Woodbridge)* 'Oll-er-t'n'
Wilford Hundred. AS. *alra tun* - 'Farmstead by the alder trees' *Alretuna* (DB). The hall stands immediately W of the church and alders probably once fringed its ancient moat. The Domesday manor of Little Cross (*Litelcros*) was probably situated in the parish.

ALDHAM *(2 m NE of Hadleigh)* 'Old-um'
Cosford Hundred. O.Merc. *alda ham* - 'The old homestead' *Aldeham* (DB). The Saxon church of St Mary is built on a mound partly surrounded by a moat of undetermined age and purpose - perhaps the site of the old homestead.

ALDRINGHAM *(5 m SE of Saxmundham)* 'Ald-en-um'
Blything Hundred. O.Merc. *Allrinces ham* - 'Allrinc's homestead' *Alrincham* (DB).

ALPHETON *(6 m N of Sudbury)* 'Al-veet-'n'
Babergh Hundred. AS. *AElfflaede tun* - 'Farmstead or village of a woman called AElfflaed' *Alflede(s)ton* (DB); *Alpheldton* (1199); *Alfeton and Alffleton* (1275).

AMPTON *(5 m N of Bury St Edmunds)* 'Amm-t'n'
Thedwestry Hundred. AS. *Amman tun* - 'Farmstead or village of a man called Amma' *Hametuna* (DB). Morley suggested that this may have been the home of Anna, the East Anglian Saxon king, and that the river had been artificially diverted here (N of Knight's Grove) to form a fortified enclosure.

ASHBOCKING *(7 m N of Ipswich)* 'Ash' also 'Ash Bork'n'
Bosmere & Claydon Hundred. AS. *aesce Boccinga* - 'Place at the ash-tree(s)' *Assa* (DB); *Bokkynge*

Assh (1411). Bocking was later added to the original name Ash following the acquisition of the manor by the de Bocking family in 1342. Their residence, Ash Hall, was situated near the church. The present hall on the same site was built in the C16.

ASHBY *(6 m NW of Lowestoft)* 'Ash-by'
Mutford & Lothingland Hundred. Norse. *Aska by* - 'Aski's building' *Assefelda* (DB). Ashby, Bradwell and Oulton are all omitted from Domesday Book assessment for this Hundred. A possible reason for this is their owner's resistance to William's Conquest for which the Crown had seized (but not regranted) their lands, as was usual in 1075. As Crown land it would not be assessable.

ASHFIELD-(CUM-THORPE) *(3 m E of Debenham)* 'Ash-vul'
Threadling Hundred. AS. *aescefeld* - 'Ash tree clearing' *Assefelda* (DB). A place where ash trees were felled, and only a field in so far as that 'felling' constituted a clearing which, when tilled, became a field. Thorpe Hall, near its ruined church, represents the hamlet of Thorpe (*Torp*).

ASPALL *(5 m S of Eye)* 'Arz-ple'
Hartismere Hundred. AS. *aesp heale* - 'Sheltered spot by the stream where aspens grow' *Aspala* (DB). The Derry Brook which separates the hall from the church indicates the place where the original aspens grew well over a thousand years ago.

ASSINGTON *(4 m ESE of Sudbury)* 'Ass-e-tun' now 'Ass-ing-t'n'
Babergh Hundred. AS. *Asan tun* - 'Assa's farm' *Asetona* (DB); *Assintona* (1175).

ATHELINGTON *(4½ m SE of Eye)* 'Arl-en-t'n' *from the 1546 spelling Arlington.*
Hoxne Hundred. AS. *oethelinga tun* - 'Farm or town of the princes or nobles' *Alinggeton* (1219). One of four Hoxne villages omitted from Domesday. The name suggests there may once have been a great house here belonging to the royalty of the East Angles with elaborate earthworks and moats. Athelney is now a hill rising conspicuously from broad surrounding marsh flats; its name perhaps originally Aethelinga, i.e. Princes Island.

BACTON *(6 m N of Stowmarket)* 'Back-t'n'
Hartismere Hundred. AS. *Bacan tun* - 'Farmstead of a man called Baca' *Bachetuna* (DB).

BADINGHAM *(4 m NNE of Framlingham)* 'Bad-'n-gum'
Hoxne Hundred. AS. *Badinga ham* - 'Bada's family Home' *Badincham* (DB). The sites of several ancient mounds are suggestive of Bada's first settlement: the church stands on the slope of a hill known as Burstonhaugh (later Derhaugh); W of the Old Rectory is another mound, and the road at Castle Farm traces a suspicious curve round a site intriguingly once named Mottshall. The site of Badingham Hall also dates back, at least, to Norman times.

BADLEY *(2 m NW of Needham Market)* 'Badd-y'
Bosmere & Claydon Hundred. AS. *Badan leah* - 'Bada's meadow' *Badeleia* (DB). The meadow may have been at the source of the Bat stream, an affluent of the river Gipping which rises at Lady Well S of the church and Badley Hall.

BADWELL ASH (formerly Little Ashfield or Ashfield Parva) *(4 m SE of Ixworth)* 'Baddle-Ash', *but more usually* 'Badd-wull-Ash'
Blackbourn Hundred. AS. *Badan wella aesce* - 'Bada's spring or well by a clearing where ash-trees were felled' *Ayshfeld Parua* (1280); *Badewelle Asfelde* (C13); *Badewell Pava* (1340).

BALLINGDON-CUM-BRUNDON *(To the W of, and now a part of Sudbury)*
Babergh Hundred. 'The hill of Baell's family' *Baldinigcotum; Belidune; Balydon; Baliton*. The A131 road S to Halstead leaves Sudbury by Ballingdon Hill, site of the ancient hall.

BARDWELL *(2½ m N of Ixworth)* 'Bard-ell' *or* 'Bard-well'
Blackbourn Hundred. AS. *Beordan wella* - 'Well of a man called Beorda' *Berdeuuella* (DB). The church stands high and overlooks the tributary marshes (doubtless all water in Saxon times) and the original 'well' or mere of the place-name.

BARHAM *(5 m NNW of Ipswich)* 'Ba-rum'
Bosmere & Claydon Hundred. O.Merc. *berh*; AS. *hamm* - 'Homestead or enclosure by the (burial) hill' *Bercham* (DB). "This town preserves an old Saxon word Barrow or Burgh being on a hill cast up. Two such barrows are openly seen: one, the biggest, hard by the turning up to Coddenham out of the road leading from Cleydon Street to Coddenham Beacon; the other, a little further on in the same field." (Tanner MS. 1818). The hamlet of Sharpstone - AS. *Sceafes tun*, includes the manor of Shrubland. Six times quoted *Scarvestuna* in DB it lies in the area of The Sorrel Horse Inn and the junction of Norwich Road and Sandy Lane (to Coddenham) now known as Sharpstone Street.

BARKING *(1 m SW of Needham Market)* 'Bar-k'n'
Bosmere & Claydon Hundred. O.Merc. *Bercingas* - 'Settlement of the family or followers of Berca' *Berchingas* (DB). At Domesday Barking had a mill and the weir of another, showing the parish then extended to the river Gipping. During at least 1388 to 1560 the Typtot and Wentworth families of Nettlestead Hall paid 'castle-guard' rent to Framlingham Castle for half a knight's fee in 'Tunstall by Nettlestead' which seems the sole later reference to the Domesday *Tonestala* (AS. *Tons steall* - 'Ton's place'). This place was not actually in Nettlestead and may refer to the ancient and much corrupted Taston Hall in Barking.

BARNARDISTON *(5 m NW of Clare)* 'Branson' *(1764)*, 'Barniston' *(1910)*, now Barn-er-diss-t'n'
Risbridge Hundred. Norm. *Bernard*; AS. *tun* - 'Bernard's manor' *Bernardeston* (1194). A hamlet or mere manor of Hundon at Domesday, it may have been known as *Wigmundston* till about 1100. *Cileburna* (DB) - Chilbourn is a hamlet in this parish which also includes the hamlet of Monks' Risbridge (Dan. *Hrisa bricge* - 'Hrisa (or Hrisi)'s bridge'), extraparochial without church, house or population and from where the Hundred took its name.

BARNBY *(4 m SE of Beccles)* 'Bar-na-by'
Mutford & Lothingland Hundred. Dan. *Barna*; Norse. *by* - 'Barni's building' *Barnebi* (DB). The church with some Saxon stone, and bricks that may be Roman, is patronised by St. John. Its situation on the Waveney south bank and near the coast is just such as a Viking would select c870 A.D. Barnby is one of only four Danish settlements in Suffolk.

BARNHAM (St Gregory & St Martin) *(9 m N of Bury St Edmunds)* 'Barn-um'
Blackbourn Hundred. AS. *beorn (Beorn) ham* - 'Warrior homestead' or 'Homestead of a man called Beorn' *Byornham* (c1000); *Bernham* (DB). The Iron-Age fort here is one of only two belonging to the period discovered in Suffolk and may identify Beorn's (or the warrior) homestead'.

BARNINGHAM *(6 m NNE of Ixworth)* 'Barn-e-gam'
Blackbourn Hundred. O.Merc. *Berninga ham* - 'Homestead or village of the family or followers

of Bern' *Bernincham* (DB). The parish boundary runs down the centre of Coney Weston village leaving half the village in Barningham.

BARROW *(6 m W of Bury St Edmunds)* 'Barr-er'
Thingoe Hundred. Mid. Engl. *berewe* from AS. *beorge* - 'Hill or Tumulus' *Baro* (DB). 'Apparently the nominal hill is not to be sought among the extensive entrenchments of Barrow Hall' (*Victoria County History*), or 'in the Saxon tumulus at Barrow Bottom that was opened in 1813' (*SI*), but in the unspecified site of that obviously Celtic "leaf-shaped sword of bronze found at Barrow."'

BARSHAM *(2 m SW of Beccles)* 'Bar-shum'
Wangford Hundred. AS. *Baeres ham* - 'Home of a man called Beare' *Barsham* (DB). The possible site is that of the moated manor house of Barsham Old Hall.

BARTON MILLS *(1 m SE of Mildenhall)* 'Bart'n Mills'
Lackford Hundred. AS. *bere tun (parva)* - 'Small corn or barley farm' *Bertuna or Bartona* (DB). The original two Bartons in the county were distinguished by being named Great Barton and Little Barton (or Barton Parva) - now Barton Mills. The latter later derived the name from its very early mills beside the river Lark.

BATTISFORD *(3 m SW of Needham Market)* 'Batts-fud', *sometimes* 'Basforth'
Bosmere & Claydon Hundred. AS. *bates forda* - 'Ford of the Bat (stream)' *Betesfort* (DB). The ford in question would appear to have been that at the bottom of Hascot Hill where the road crosses the Bat stream rising from the ancient Lady Well in the adjoining parish of Badley.

BAWDSEY *(10 m SE of Woodbridge)* 'Bord-zy'
Wilford Hundred. O.Merc. *Baldheres*; AS *ig* - 'Island or promontory of a man called Baldhere' *Baldereseia* (DB). The place-name indicates that, until sea levels dropped, this coastal parish may once have been an island, or at least a promontory, with Bawdsey Marshes to the S and Alderton Marshes to the W and N. Bawdsey includes the former smuggling haunt of Shingle Street.

BAYLHAM *(3 m SSE of Needham Market)* 'Ba-lum' *pronounced as in 'bay'*
Bosmere & Claydon Hundred. O.Icel. *boeli*; AS. *hamm* - 'Farm enclosure' *Beleham* (DB). Baylham Hall retains a part of its former strong moat and may be the site of the farm enclosure.

BECCLES 'Beck-uls'
Wangford Hundred. AS. *Beccles* - 'Beccel's place' or 'Place by the stream in a pasture' *Becles* (DB). The place-name may refer to a tributary of the Waveney to the west of the town. Situated on a bold promontory above the river, this was an important town at Domesday when the river could accommodate large herring fishing fleets using the quays here. It is interesting to note that the streets are named from the Norse word *gata* e.g. Blyburgate, Hungate, Smallgate, etc.

BEDFIELD *(4 m NW of Framlingham)* 'Bay-a-fel' *(very early)*, 'Bedd-feld'
Hoxne Hundred. AS. *Bedan feld.* - 'Clearing of a man called Beda' *Berdefelda* (DB).

BEDINGFIELD *(4 m SE of Eye)* 'Bedd-'n-feld'
Hoxne Hundred. AS. *Bedinga feld* - 'Clearing of the family or followers of a man called Beda' *Bedingefelda* (DB).

BELSTEAD (formerly Belstead Parva) *(4 m SW of Ipswich)* 'Bell-stid'
Samford Hundred. ?Norse. *Belta*; AS. *stede* - Beli's place' or 'Place in a glade' or 'place of a funeral pyre' *Belesteda* (DB). Formerly divided into Magna and Parva, with Belstead Magna

later becoming Washbrook. Any of the possible place-name meaning could be applicable here as Belstead Brook runs through the parish, while the area's light sandy soil provides conditions for a former Bronze-Age burial mound. Two gold torques were found N of Belstead Brook and the parish boundary in the 1960s. Stone Lodge Lane is an ancient way which crossed Belstead Brook just below Gusford Hall at the Domesday *Gutthulvesforda* - 'Guthwulf's Ford' in the parish of St. Mary, Stoke in Ipswich but whose manor appears to lie mainly in Belstead.

BELTON *(5 m SW of Yarmouth)* 'Bell-t'n'
Mutford & Lothingland Hundred. ?Norse. *Beja*; AS. *tun* - Beli's farm' or 'Place in a glade' or 'place of a funeral pyre' *Becketuna & Bechetuna* (DB). Roman pottery has been found in a round barrow at Mill Hill S of the village which may have given the parish its name.

BENACRE *(7 m SE of Beccles)* 'Benn-er-ker'
Blything Hundred. AS. *bean aecer* - 'Cultivated plot where beans are grown' *Benagra* (DB).

BENHALL *(1½ m SW of Saxmundham)* 'Benn-ull'
Plomesgate Hundred. AS. *Beonan heale* - 'Beona's shelter' *Benenhala* (DB). Benhall Place, a sheltered spot on the E bank of the river Alde, is the likely early settlement site. Friday Street is a Pagan Saxon name and Kelton, *Chiletuna* (DB) lay W of the church and formerly had a market.

BENTLEY *(5½ m SW of Ipswich)* 'Bent-ly'
Samford Hundred. AS. *beonet leah* - 'Coarse-grass meadow' *Benetleia* (DB). Two other Domesday townships appear to lie here: *Mamesfort* (now lost) and Dodnash AS. *Dudam aesce* - 'Duda's ash-tree' *Dudenasch* (DB). The ash was a sacred tree in Pagan Saxon times and Dodnash monastery, founded in 1188, once stood here.

BEYTON *(5½ m SE of Bury St Edmunds)* 'Bay-t'n'
Thedwestry Hundred. AS. *Baegan tun* - 'Farmstead of a woman called Beage or a man called Baega'? *Begatona* (DB).

BILDESTON *(5 m NW of Hadleigh)* 'Bill-sten' *or* 'Bilston' *with no trace of 'd'*
Cosford Hundred. AS. *Byliges tun* - 'Farmstead (or hill/promontory) of a man called Bylig' *Bilestuna* (DB). The church and site of the former hall, standing high and detached some distance from the present village, form the nucleus of the original settlement.

BLAXHALL *(4 m SW of Saxmundham)* 'Blacks-el'
Plomesgate Hundred. AS. *Blaeces heale* - 'Blaec (Black)'s shelter' *Blaccheshala* (DB). The original settlement sheltered beside the river Alde (perhaps near the present hall) and not on the high ground where the present church stands.

BLUNDESTON *(3 m NW of Lowestoft)* 'Bluns-t'n'
Mutford & Lothingland Hundred. Norse. *Blundes*; AS. *tun* - 'Blundr's farmstead' *Blundeston* (1203). Blundeston Hall, SE of the church, still retains part of an ancient moat which once surrounded an earlier hall.

BLYFORD (or Blythford) *(2½ m SE of Halesworth)* 'Bli-fud'
Blything Hundred. AS. *Blithe forda* - 'Ford of the river Blyth (meaning gentle or pleasant one)' *Blideforda* (DB). The ford was probably sited at Wenhaston Bridge.

BLYTHBURGH *(4½ m W of Southwold)* 'Bli-brah'
Blything Hundred. AS. *Blithe burh* - 'Town (fortified place) on the river Blyth' *Blideburh* (DB).

While there is conjecture over Blythburgh's early origins, there is no doubt that it was a well established *burh* (town) in Saxon times when it was a flourishing port with a considerable fishery and its own market. Its present hamlets were distinct townships in 1086: Bulcamp, L. *Belli campus* - 'Field of battle' *Bulecampa* (DB), was the site of the battle between the Saxon King Anna and the Danish King Penda in 654 AD; Hinton, AS. *hina* - 'agricultural labourers' *Hinetuna* (DB), now represented by Hinton Hall; and the Domesday *Bringas*, situated close to the bridge on the Wenhaston boundary. The assignment and definite site of the Domesday township of *Alnetuna*, *Ernetuna* or *Ernetherne* (c1090) is uncertain.

BOTESDALE *(7 m NW of Eye)* 'Boods-el', 'Budds-dle', Buddesdale' *or* 'Bod-es-dle'
Hartismere Hundred. AS. *Botwulfes dale* - 'Valley of a man called Botulf' *Botholuesdal* (1275). Always assumed as the birthplace of Botwulf who built the monastery at Iken in 654 A.D. and died in 680. Only two facts go to substantiate this claim: the adjacent village of Redgrave was granted a fair in 1227 upon St. Botulf's Day which is that of his death on 17 June, and the Chapel-of-ease in the main street was found to have been dedicated to the same saint upon the occasion of a Chantry's foundation therein in the C15 - of the chapel's own foundation no record survives and its position high on the Dale's east bank is not such as one usually regards as so early as *c*680 AD.

BOULGE *(2½ m N of Woodbridge)* 'Bowge', *also* 'Bow-idge'
Wilford Hundred. O.Norm. *Bulges* - 'Heather-covered waste' *Bulges* (DB). The parish is always associated with Debach and doubtless formed a part of it in early days. Odo de Campania held *Bulges* and *Depebeck* at Domesday, and Sir Thomas Hanmer was later lord of both. It is likely at this time that Debach was under forest and Boulge, though 'heather-covered waste', the area of any settlement.

BOXFORD *(5 m SW of Hadleigh)* 'Box-ford'
Babergh Hundred. AS. *box forda* - 'Ford by the box-trees' *Boxford* (C12). The church stands by the crossing of the river Box in the centre of the village, site of the ancient ford, and no doubt surrounded by box trees.

BOXTED (formerly Boxstead) *(6 m NE of Clare)* 'Box-id'
Babergh Hundred. AS. *box* or *boc steda* - 'Place where beech-trees or box-trees grow' *Boesteda* (DB); *Boxted* (from 1275). The moated Boxted Hall is situated in the valley below the church.

BOYTON *(7½ m SE of Woodbridge)* 'Boy-t'n'
Wilford Hundred. ?AS. *Boian tun* - 'Farmstead of a man called Boga' *Boituna* (DB).

BRADFIELD *(5 m SE of Bury St Edmunds)* Bradd-feld S'n Jorge, Kum-Bust, S'n Klah'
Thedwestry Hundred. AS. *bradan feld* - 'Broad clearing' *Bradfelda* (DB). A single township until Domesday, its sole church in what is now Bradfield St. George. It was given to Bury Abbey and became Monks' Bradfield. The second manor was also given to the Abbey and its later grange burned in the Peasants' Revolt of 1329 when a subsidy termed it Parva Bradfield, later Brent Bradfield and later still Bradfield Combust - the *Victoria County History* states that it was also known as Bradfield Manger, and there is a public house of that name in the village. The manorial chapel of the third, founded by the Seyncler family, emerged in 1317 'ascribed to Saint Clare'.

BRADWELL *(8 m NW of Lowestoft)* 'Bradd-wee'
Mutford & Lothingland Hundred. AS. *bradan wellan* - 'Place at the broad stream or lake' *Bradewell* (1211). The hall lies close to the southern shore of the erstwhile estuary of the river

Yare, far from its upland church, and the 'broad stream' no doubt refers to the estuary. The Lothingland township of *Hornes* is perhaps represented by Hobland Hall which occupies a possible Danish site.

BRAISEWORTH *(2 m SW of Eye)* 'Brazer'
Hartismere Hundred. AS. *Brysan worth* - 'Brysa (Bruiser)'s property' or 'Enclosure infested with gadflies' *Briseworde* (DB). Of the old church here no more than a portion of the chancel remains and that in a semi-ruinous condition. It is enough to definitely fix the very early site upon a distinct hoh only some 20' above the marshes now representing the old level of the river Dove.

BRAMFIELD *(2 m S of Halesworth)* 'Brarm-feld', *or now* 'Bramm-field'
Blything Hundred. ?AS. *brom feld* - 'Broom clearing' *Brunfelda* (DB).

BRAMFORD *(2½ m NW of Ipswich)* 'Brarm-fudd'
Bosmere & Claydon Hundred. ?Prov. Engl. *brame*; AS. *forda* - 'Ford by the bramble' *Branfort* (DB). Includes *Runcton* ('Runnton'), a former hamlet N of Sproughton and represented by Runcton Farm. The B1067 road crosses the river Gipping near the church, site of the old ford.

BRAMPTON *(5 m NE of Halesworth)* 'Bramm-t'n'
Blything Hundred. AS. *brom tun* - 'Farmstead where broom grows' *Bramtuna* (DB). Brampton Old Hall stands on an ancient moated site.

BRANDESTON *(4 m SW of Framlingham)* 'Brann-es-t'n'
Loes Hundred. AS. *Brandes tun* - 'Farmstead of a man called Brand' *Brantestona* (DB). Friday Street, a collection of cottages 10' above the river Deben and upon the extreme edge of its ancient level, is Pagan Saxon in origin, older than the name of the village.

BRANDON *(4 m NE of Mildenhall)* 'Brann-d'n'
Lackford Hundred. AS. *brom dun* - 'Broom hill' *Bromdune* (Late C10), *Brandona* (DB). A tumulus in Brandon rises upon the highest point of Ling Heath on a spot now hidden in pinewood ½ mile from the town. As late as 1250 a fen extended from Ely to Thetford passing Brandon.

BRANTHAM *(8 m SW of Ipswich)* 'Brann-thum'
Samford Hundred. AS. *brant hamm* - 'Steep enclosure' *Brantham* (DB). Brantham Hall doubtless marks the original settlement site at the head of the gully running up from the London-clay splay of the river Stour at Seafield Bay. The parish includes the hamlet of Cattawade - 'Kati's ford' or O.Norse. *kata* meaning a kind of small boat or ship. If the latter, it would have applied to a deeper boat-ferry rather than a ford.

BREDFIELD *(3 m N of Woodbridge)* 'Bred-feld'
Wilford Hundred. O. Frisian. *breed fold* - 'Broad clearing' *Bredefelda* (DB).

BRENT ELEIGH *(2 m SE of Lavenham)* 'Brent-e-ly'
Babergh Hundred. AS. *Illan leah*. 'Illa's meadow' *Munegadena* (DB); *Monekesillegh* (1304). This was one single entity till the year 991, and Illa was most likely a monk as the manor was given to the monks of Canterbury by Brithnoth, Earl of Essex, who lost his life in 991 A.D. fighting the Danes in the Battle of Maldon. It was then called Monks Eleigh. After the Dissolution it was given to the Dean and Chapter of Canterbury. A part, Brent Eleigh, then became detached from the main parish and gained its own identity. The name, meaning 'burnt', appears to have replaced Monks when a fire destroyed the former village in 1254. It was called Illea Combusta by Copinger and Illeghe Combust in 1327.

BRETTENHAM *(5 m NE of Lavenham)* 'Brett-num'
Cosford Hundred. AS. *Bretta ham* - 'Homestead of a man called Bretta' *Bretenhama* (DB). East of the Hall rises the river Brett. The former Domesday hamlet of *Rushbrooke* in this Hundred is obviously distinct from the village of that name in Thedwastre Hundred. The hamlet is now lost but was evidently associated with Brettenham and probably formed a part of it.

BRIGHTWELL *(5½ m SE of Ipswich)* 'Brit-wull'
Bosmere & Claydon Hundred. ?AS. *beorhtan wellan* - 'At the clear well' *Brihtewella* (DB). Springs from which the parish no doubt gained its name lie to the SE of the church and hall.

BROCKLEY *(6½ m SW of Bury St Edmunds)* 'Brock-lee'
Thingoe Hundred. AS. *broc leah* - 'Brook meadow' *Broclega* (DB). Brockley Hall and church stand on the bank of the Chad Brook and provide the likely initial settlement site.

BROME *(2 m N of Eye)* 'Broom'
Hartismere Hundred. AS. *brom* - 'Place where broom? grows in abundance' *Brom* (DB). Morley states, however, that with Brome lying entirely upon heavy boulder clay, broom, would never have flourished or even grown here. He suggests the species meant was more likely AS. *gorst* - 'gorse'.

BROMESWELL *(2 m NE of Woodbridge)* 'Brumms-ell' *or* 'Brumswall'
Wilford Hundred. AS. *Brumes wella* - Brum's well' *Bromeswella* (DB). Here we find arid soil and a major part of the area still heathland; so the well was near or on the river Deben, possibly at the present Sink Farm or covering the marshy 'Common'. Arnott places the Domesday *Thurstantestuna* (Thurstanton) in this parish, N of Little Hill.

BRUISYARD *(4 m NE of Framlingham)* 'Broose-yudd'
Plomesgate Hundred. O.Merc. *burgwardes*; AS. *geard* - 'Burhward's enclosure' *Buresiart* (DB). The only Suffolk village (with the exception of the long lost Plumyard in Trimley) with the ending -*geard*, which is the Saxon word for an enclosed yard. Morley states 'I do not consider this a proper name but the ward or guardian of the *burh* i.e. fort; and follow the hypothesis that this fort was Framlingham Castle of which we know nothing pre-Conquestial. There is no historic evidence to support the traditions that a Saxon burh occupied the castle site but Bruisyard lies concealed in a gully just three miles NE of Framlingham, in a line with Dunwich; and the presence of such an outpost, in exactly the direction whence the Norsemen would attack the castle, is strong presumptive proof of the latters existence long ere 1066.'

BRUNDISH *(4½ m N of Framlingham)* 'Brunn-dish'
Hoxne Hundred. AS. *burn edisc* - 'Enclosed pasture on a stream' *Burnedich* (DB). The stream referred to is the river Alde which rises at Brundish Church (with its typically Saxon E tower window) and at this point near its source, only a stream.

BUCKLESHAM *(6 m SE of Ipswich)* 'Buck-el-shum'
Colneis Hundred. AS. *Bucles ham* - 'Homestead of a man called Bucel' *Bukelesham* (DB). Anciently known as Bulechamp, Bucklesham was the C14 demesne of William de Kerdeston. It includes the Domesday township of *Kenbroke* (now represented by Kembroke Hall in the S of the parish) with its church of All Saints, which may have been sited in Chapel Road running W from the hall. Mill River, which rises in Foxhall and terminates at Kirton Creek on the Orwell, was likely called *Kenebroc* in 1086 from the AS. *cene broc* (with long *o*) - 'the turbulent brook', from a time when rivers were likely some 5' higher than today and presented a very different picture.

BUNGAY 'Bung-y'
Wangford Hundred. Icel. *bungi*; Angl. *eg* - Elevated island' or AS. *Buna. inga. eg* - 'Island of the family or followers of a man called Buna' *Bunghea* (DB). The river Waveney forms a loop around Outney Common virtually making an island (and formerly probably did) of the N of the town. The higher ground thus enclosed and protected constituted a capital strategic position which the Vikings termed *Bung* from the Icelandic *bungi* and Norwegian *bunga*, meaning an elevation, closely allied to the Danish *bunke* and the common English word bunch. The original sense was 'rounded elevation'. This places the name's origin c870 AD at the earliest, but that it was settled earlier still is suggested by the fact that 'a silver penny of Offa, King of Mercia (757-796 AD) was dug up in the osier ground near the castle ramparts in 1813.'

BURES ST MARY *(5½ m SE of Sudbury)* 'Bew-ers' *as in 'sewers'*
Babergh Hundred. O. Norm. *bures* - 'A group of dwellings or cottages' plus church dedication. *Bura* (DB). Sometimes called Bures St Mary in distinction from Mount Bures just across the border in Essex. The parish church is on its early site upon the alluvium of the river Stour's former bed.

BURGATE *(5 m NW of Eye)* 'Burr-gutt'
Hartismere Hundred. AS. *burh geat* - 'Fort barrier or outpost' *Burgata* (DB). Though unlikely, it has also been suggested to mean 'Burial Gate' where the body of King Edmund lay for one night on its journey to Bury Abbey. From the Saxon *geat* 'moveable barrier' and not the Norse *gata* meaning street. The Saxon *burh* primarily indicates a fortress and 'I am of the opinion that the now lost place Burghgate near Felixstowe was a barrier outpost of Walton Castle' (Arnott) just as Burhward's Yard (Bruisyard) was one of Framlingham Castle. No less here perhaps Burhgate was a barrier outpost of Eye Castle and may be represented by the earthworks in Burgate Wood W of the church or in the moated site of the old hall.

BURGH *(3½ m NW of Woodbride)* 'Burg' *as in ice-'berg*
Carlford Hundred. AS. *burh* - 'Fortified place' *Burc* (DB). The original Saxon *burh* meant a fortress, a defensible stronghold. The three Suffolk Burghs had all been Roman forts before the Saxons termed them. The present village lies on the military road from Stratford St Mary to (probably) Dunwich. The site of the place-name is situated in Castle Field beside the road extending either side of Drabs Lane. An important Belgic Iron-Age fortified site belonging to the Trinovantian tribe from which much fine pottery has been obtained, is overlayed in the northern corner by an early Roman settlement. Thistleton here constituted the Domesday hamlet of *Thisteldena* or 'Thistly Valley'.

BURGH CASTLE *(4 m SW of Great Yarmouth)* 'Burr-rah Cass-ell'
Mutford & Lothingland Hundred. AS. *burh*; Mid-Engl. *castel* - 'Fortress - fortress' *Burch* (DB). This foolishly tautological name arises from medieval folk forgetting that 'burh' meant 'fortress' and, consequently, tagging on the same word in a later laguage. Situated above the marshy ground where the river Waveney joins Breydon Water, the spectacular remains of *Gariannonum*, a Roman Fort built around 290 A.D. and garrisoned by a body of Cavalry (the Stablesian horse), formed one of a series built along the east coast under the command of the Count of the Saxon Shore to defend this outpost of the Roman Empire against attack from northern invaders. The Saxon's termed it *Cnobhere's Burh* "from Cnobhere, a Saxon chief who formerly resided here."

BURSTALL *(4 m W of Ipswich)* 'Bus-tle' or 'Burs-tle'
Samford Hundred. AS. *burg steall* - 'Site of a fort or stronghold' *Burgestala* (DB). Probably the stocks or stones of a ruinous fortress were then still actually visible. Though no site has been identified in connection with the place-name, the village may have stood 'at the site of' (where nothing remained, even then) a Roman (or earlier) earthworks unrecorded in history and now long lost and forgotten. Interestingly, there is a Fort Cottage, right on the parish boundary in an elevated position beside the road SE of the church and village.

BURY ST EDMUNDS 'Bury' *(as in the verb to 'bury')*
AS. *Sancte Eadmundes byrig* - 'At St. Edmund's town'. The Saxon settlement of *Beodericeworth* (985 AD) became the fortified town of *Byrig* in 1038, and Bury St Edmunds in 1065 after the C9 East Anglian King Edmund, who was interred and later canonised here.

BUTLEY *(7 m E of Woodbride)* 'Butt-ly'
Loes Hundred. Norse. *Butta;* AS. *leah* - 'Woodland clearing of a man called Butta' *Butelea* (DB). Its Norse origin would seem to date this place-name to *c*900 AD. A small tributary of the river Ore, rising in Office Wood in Eyke, takes name from this village as is obvious from its usual designation 'The Butley River'; though in 1735 Kirby terms it 'Puddle Water'. Staverton, formerly a Domesday vill but now a part of Butley and Eyke, is from the AS. *stearfa*, meaning 'slaughter'. The ancient Staverton Forest contains an earthworks which has been identified as a possible former deer pound - perhaps where they were culled.

BUXHALL *(3½ m SW of Stowmarket)* 'Bucks-ull'
Stow Hundred. AS. *Bucces healh* - 'Bucc(Buck)'s shelter' *Buckeshala* (DB). Literally 'of a buck' employed as a personal name. An obvious association with Bucc's shelter would have been the ancient site of Fasbourn Hall which appears to originate from *Facheduna* except that this Domesday vill is located in the adjoining Bosmere Hundred.

CAMPSEY ASH *(2 m E of Wickham Market)* 'Kamps-ey-ash' *or just* 'Ash'
Loes Hundred. ?Norse. *kampes*; AS. *eg. Aesce* - 'At the ash near Campi's promontory'. *Campeseia* (DB). *Esce* (DB). Two quite distinct settlements at Domesday. The original ash-tree doubtless grew at the island in the middle of a splay of the river Deben whereupon the nunnery of Ash Abbey was founded c1195, and may represent the site of the first hall. Campsey does not refer to this 'sea' or river expansion E of Wickham Market, but a tongue of upland (*eg*) jutting out into it, possibly at the present Quill Farm.

CAPEL ST ANDREW *(7 m E of Woodbride)* 'Kay-ple'
Wilford Hundred. O.Norm. *capeles* - 'Chapel' plus dedication. *Capeles* (DB). A very late place-name. It suggests an ancient chapel in the parish dedicated to St Andrew, but this may have been a reference to the Saxon church mentioned in Domesday. A church with this dedication later given to Butley Priory existed here in 1529. Its site may have been near Home (or Hall) Farm where many human bones have been dug up.

CAPEL ST MARY *(5½ m SE of Hadleigh)* 'Kap-ple'
Samford Hundred. O.Norm. *capele* - 'Chapel' plus dedication. *Capeles* (DB). Again a very late place-name not applied to the village until sometime between 1086 and 1275. One of the three manors here is called Churchford and is situated in the N of the parish through which a brook flows; it is represented by Churchford Hall and Churchford Farm. It comes from the AS. *circe ford*

and comprised the distinct township of *Cercesfort* at Domesday. This may be a clue to the site of the chapel.

CARLTON *(1 m N of Saxmundham)* 'Karl-t'n'
Hoxne Hundred. AS. *carla tun* - 'Farmstead or estate of the freemen' *Carletuna* (DB). Not a true Saxon word but adapted in Saxon times from the Norse word *karl*. The Scandinavian influence indicates a late date for settlement, i.e. certainly after 870 AD.

CARLTON COLVILLE *(3½ m SW of Lowestoft)* 'Kall-t'n Kall-vul'
Mutford & Lothingland Hundred. AS. *carla tun* - 'Farmstead or estate of the freemen' *Carletuna* (DB); *Carleton Colvile* (1346). It was once known as *Est Carlton* (East Carlton) but became Carlton Colville many years after the acquisition of the manor by the de Colevill family in the C13. Beside the Bell Inn is an almost complete circle of water known as The Mardles, a name applied to the spot where gathered hemp is placed to soak. From the surrounding springs rose Kirley Run stream.

CAVENDISH *(6 m NW of Sudbury)* 'Ka-v'n-ish' *or* 'Candishe'
Babergh Hundred. AS. *Cafan edisc* - 'Cafa's late-mown land' *Kauanadisc* (DB). The modern spelling *Cavendyssh* does not appear until 1381. Excavations in 1983 on earthworks in a pasture covering 3.6ha. identified hollow-ways, banks and ditches and an oval enclosure. An C11 date has been suggested for the enclosure which may possibly identify the original Cafa's Hall and its associated settlement.

CAVENHAM *(4½ m SE of Mildenhall)* 'Kav'num', *or for shortness called* 'Canham' *(Kirby 1764).*
Lackford Hundred. AS. *Cafan ham* - 'Cafa's homestead/enclosure/water-meadow' *Kanauaham* (DB); *Cauenham* (1198).

CHARSFIELD *(5½ m N of Woodbridge)* 'Chars-vul *(rare) or* 'Chars-field'
Loes Hundred. AS. *Cerres feld* - 'Cerr's clearing' *Ceresfelda* (DB). There seems to have formerly existed a village or hamlet of Charsfield Parva.

CHATTISHAM *(5 m E of Hadleigh)* 'Chatt-es-um'
Samford Hundred. AS. *Ceattes ham* - 'Ceatt's homestead or water-meadow' *Cetessam* (DB). The Hall stands high but Spring Brook cuts across the E of the parish in a shallow valley.

CHEDBURGH *(7 m SW of Bury St Edmunds)* 'Chedd-brah' *or* 'Chedbar'
Risbridge Hundred. AS. *Ceddes bergh* - 'Hill (tumulus?) of a man called Cedd' *Cedeberia* (DB). Norman Scarfe considers it may refer to a mound connected with Cedd, the founder of Christianity in Essex, who died on 26 October 664. Cedd was brother to Bishop Ceadda of Lichfield (669-72). Chedburgh Hall, now a farmhouse near the church, stands on the site of the ancient manor house and still retains traces of an earlier moat.

CHEDISTON *(2 m W of Halesworth)* 'Chedd-es-t'n' *or* 'Cheston'
Blything Hundred. AS. *Ceddes stan* - 'Stone of a man called Cedd' *Cedestan* (DB). Ironically, whereas in most instances the *e* has been erroneously added to many Suffolk place-names, in this instance, where it is relevent, it has been dropped. The name should be Chedistone, and the erratic from which it took its name lies near Rockstone Manor Farm in neighbouring Cookley. Morley states that the (sand)stone measuring 10'x6' lies near Grange Farm in Rockstone Lane. Saint Cedd may have preached beside it around 660 AD.

CHELMONDISTON *(6 m SSE of Ipswich)* 'Chempton' *(early)*, Chimpton or 'Chel-merdeson'.
Samford Hundred. AS. *Ceolmundes tun* - 'Farmstead of a man called Ceolmund' *Canapetuna* (DB); *Chelmundeston* (1174). It is likely that the Saxons called it Campton from where the local colloquialism Chempton appears. The Domesday *Alsildestuna* is also placed here with less authority as no lady of this name emerges in Suffolk history.

CHELSWORTH *(5 m NW of Hadleigh)* 'Chells-uth'
Cosford Hundred. AS. *ceorles worth* - 'Husbandman's property' *Ceorleswyrthe* (962); *Cerleswrda* (DB). Joined here by two tributary streams, the Brett skirts the rising ground called Park Fields, now Chelsworth Park. The present hall is modern but traces of the ancient Hall were said to have been seen within the park in the C19 on an island site near the Grange with the river forming the S side. Here, C16 court rolls talk of 'Le Mote and Hall Gardens', and a description in 1605 states 'No mention where the mannour house stoode. Yet at the W end of the churche appeareth the seate of a most ancient house environed with ditches and moates . . .'

CHEVINGTON *(6 m SW of Bury St Edmunds)* 'Chev-en-d'n.'
Thingoe Hundred. AS. *Cweofan tun* - 'Ceofa's estate/farmstead' *Ceuentuna* (DB). The intrusive *g* does not appear in the place-name until the 1327 Subsidy. The present C17 Chevington Hall, standing on the N side of the church with its massive horse-shoe moat, occupies an exceedingly strong, ramparted and irregular ancient moated camp, which was later the site of the fortified manor house of the abbots of Bury. The site, which may well be that described in the place-name, formed part of the estate of Britwulf the Saxon, conferred on Bury Abbey at the Conquest at the express wish of Abbot Baldwin. It later became a favourite retreat of the abbots who used the earthworks to protect the country house they built within which was mentioned in 1309.

CHILLESFORD *(9 m NE of Woodbridge)* 'Chill-fud'
Plomesgate Hundred. Mid. Engl. *chisel forda* - 'Gravelly ford' *Cesefortda* (DB). The original ford was probably at the northern extremity of the Butley River just S of the church, from where 'the peculiar stratum Chillesford Crag' was found in 1865.

CHILTON *(1 m NE of Sudbury)* 'Chill-t'n'
Babergh Hundred. AS. *Ceolan tun* - 'Ceola's farm' *Ciltona* (DB). There was a hall here at Domesday, probably on the same site as Ceola's farm referred to in the place-name. The present isolated Chilton Hall is situated in high clay lands NE of the church and some ¼ mile from the Waldingfield road. It stands on the C13 site of an earlier fortified manor house and was long the seat and stronghold of the powerful knightly family of Crane, one of whom, Sir Robert, lies in a magnificent tomb in the nearby church.

CLARE 'Klay-r'
Risbridge Hundred. L. *clarus* - 'The Clear (stream)' *Clara* (DB). Clare lies on the Clarus, a tributary which joins the river Stour here. It was, perhaps, an old river-name of Celtic origin.

CLAYDON *(4 m NW of Ipswich)* 'Klay-d'n'
Bosmere & Claydon Hundred. AS. *clayen dun* - 'Clayey hill' *Clainduna* (DB). Claydon Hall, now a farmhouse SE of the church, stands within the northern side of what remains of a strong moat. The 1926 OS map shows Claydon Hall 'on the site of a castle' and it is this description of the earliest dwelling here that pervades from all early authorities. If it was indeed a castle, or at least (from the obvious former strength of the earthworks) a fortified

manor house, it could well have been the seat of an Adam Aula de Cleiden who held the manor in the middle of the C13 and occupied the site of the first settlement.

CLOPTON *(4 m NW of Woodbridge)* 'Klopp-n' *(early)*, 'Klopp-t'n' *(later)*
Carlford Hundred. ?AS. *Cloppan tun* - 'Cloppa's farm' *Clopetuna* (DB). Domesday lists a carucate of land (120 acres) in this Carlford Hundred called *Kingeslanda* i.e. 'Kings Land'. There is a Kingshall Manor in Clopton so it is likely to have been situated here.

COCKFIELD *(6 m SSE of Bury St Edmunds)* 'Cock-full'
Babergh Hundred. AS. *Coccan feld* - 'Cocca's clearing' *Cochanfelda* (DB).

CODDENHAM *(3 m ESE of Needham Market)* 'Kodd-num'
Bosmere & Claydon Hundred. AS. *Codan ham* - 'Cod(d)a's homestead' *Codenham* (DB). One of the few Suffolk examples of a Saxon name which has come down to us from Domesday (apart from the addition of another *d*) unchanged. At Domesday *Uledana* was a separate township here last mentioned in 1650 as Olden. It may have occupied the area now named Coddenham Green.

COMBS *(1 m S of Stowmarket)* 'Kooms', *pronounced abruptly as in 'brooms'*
Stow Hundred. AS. *cambas* - 'The hill-crests or ridges' *Cambas* (DB). Refers to the tops of a crest or summit such as Poplar Hill and Burntmill Hill especially those overlooking the Gipping valley up which the first Saxons would have approached the place. "Its name corresponds exactly with the parish which is a series of rising crests of land" (Hollingsworth).

CONEY WESTON *(6 m NE of Ixworth)* 'Kunn-e-wez-t'n', *u pronounced as o in 'Wholely'*
Blackbourn Hundred. O.Norse. *konungs*; AS. *tun*. - 'The king's manor - royal estate' *Cunegestuna* (DB). Its name suggests that it was one of the earliest royal estates in Suffolk, perhaps preceding the creation of the Hundreds.

COOKLEY *(2½ m WSW of Halesworth)* 'Kook-ly'
Blything Hundred. AS. *Cucan leah* - 'Clearing of a man called Cuca' *Cokelei* (DB). We may consider Cuca, after sailing past Blythburgh, more likely to have squatted alongside the main river than by the smaller tributary near the present church as suggested by Suckling - "this parish derives its name from its situation on a little stream of water or cockey".

COPDOCK *(3 m SW of Ipswich)* 'Karp-dook'
Samford Hundred. AS. *copped ac* - 'Pollarded oak-tree' *Coppedoc* (1195). The place name seems to refer to a single oak tree obviously of some stature even though its top had been removed. Morley stated that 'the name is such good and simple Saxon that we cannot suppose it was conferred after the Conquest and one fails to explain the township's total absence from Domesday Book, for it is situated on the main Roman road from Colchester to Coddenham so it must have been at least partly in hidage or arable at that very late date. Nor has this primitive title been substituted for an earlier name.'

CORTON *(3 m N of Lowestoft)* 'Kor-t'n'
Mutford & Lothingland Hundred. ?AS. *Coran tun* - 'Farmstead of a man called Cora' *Karetuna* (DB). There is doubt concerning the meaning here encouraging other interpretations e.g. "the Celtic *caer-dun* - 'a fortress-hill', was Latinised *Cor-tona* by the Romans." (Morley) Any such fortress along this stretch of coast would be long lost to coastal erosion. The hamlet of Newton, AS. *nuve tun* - 'New farm' *Neutuna* (DB), which formerly stood eastwards of Corton has been lost to the sea.

COTTON *(6½ m N of Stowmarket)* 'Kott'n'
Hartismere Hundred. AS. *Codan tun* - 'Codda's farmstead' *Codestuna* (DB). Cotton Hall lies to the S and its ancient moat is one of a number in the parish.

COVEHITHE (formerly North Hales) *(4½ m NNE of Southwold)* 'Koove-hithe'
Blything Hundred. AS. *cofa hyth* - 'Cove/landing place' *Coveheith* (1523). Covehithe forms what remains of the former medieval port of North Hales, now lost to the sea, and was one-time its probable hamlet.

COWLINGE *(8 m NNW of Clare)* 'Kull-inge', *surviving in speech in its Domesday form*
Risbridge Hundred. AS. *Culinge* - 'Of the family of Cul' *Culinge* (DB).

CRANSFORD *(3 m ENE of Framlingham)* 'Krarns-fud'
Plomesgate Hundred. AS. *cranes forda* - 'Ford frequented by cranes or herons' *Craneforda* (DB). Morley states that animal names prefixed to fords roughly indicate their respective depths at that time. Hence the river Alde here would have been too deep, and guesses the original ford to be now represented by a dip to a brook across the road to Rendham E of West Farm.

CRATFIELD *(6½ m WSW of Halesworth)* 'Kraat-vul'
Blything Hundred. Dan. *krat*; AS. *feld* - 'Thicket-clearing' *Cratafelda* (DB). The sense is actually a field which has gone out of cultivation and is now covered with brambles.

CREETING ST MARY *(2 m NE of Needham Market)* 'Kreet-en Sun Mary', etc.
Bosmere & Claydon Hundred. AS. *Crettinga* - 'Settlement of the family of Cretta' plus church dedication. *Cratingas* (DB). Once a single parish it became Creeting Magna and Creeting Parva in 1327 and later embraced the parishes of St. Mary, All Saints and St. Olaves. Bosmere, the deep lake (or mere) from where the Hundred takes its name, lies on the southern parish boundary. The extensive Newton manor *Niwetuna* (DB) comprised all the lands which lay W of the river Gipping, an area now encompassed by the parishes of Needham Market and Barking.

CREETING ST PETER (Formerly West Creeting) *(2½ m ESE of Stowmarket)*
Stow Hundred. AS. *Crettinga* - 'Settlement of the family of Cretta' *Cratingas* (DB). Although adjacent to the three other original Creeting parishes, St Peter is situated in a different Hundred. West Creeting, which first appears as *Westcretyng* in the 1381 poll-tax, alone stands in Stow Hundred.

CRETINGHAM *(4 m WSW of Framlingham)* 'Kritt-num' *or* 'Kreet-num'
Loes Hundred. AS. *Crettinga ham* - 'Cretta's family home' *Gretingaham* (DB). Formerly comprised the two parishes of Great and Little Cretingham. Unexcavated earthworks about one mile S of the village on the E of the Otley-Cretingham road may form the site of the original settlement.

CROWFIELD *(1½ m ENE of Needham Market)* 'Croo-field', *later often* 'fel' *or* 'ful'
Bosmere & Claydon Hundred. AS. *croft-feld* - 'A small enclosure or clearing' *Crofelda* (DB). Crowfield is situated at the head of a valley. The Domesday *Horswalda* (Horswold) - 'Horse Wood' is generally ascribed to Crowfield.

CULFORD *(4 m NNW of Bury St Edmunds)* 'Kull-fer'
Blackbourn Hundred. AS. *culan forda* - 'Ford at the hollow' *Culeforda* (DB). The site of the original Culford village lies near the church in Culford Park. The ford in question may have been replaced by the country lane which bridged the river Lark E of Mill Farm and

the site of the Roman settlement in Hengrave. Chimney Mills, formerly a separate parish on the southern parish boundary by the river Lark, has twice been annexed to Culford, the final time in 1897. One property plus the mill on the river - later operated by steam as well as water - were all that survived of the former parish by the 1930s. Chimney Mills and Chimney Street come from the medieval latin word *Chimnagium* which was a toll charged for wayfarage through forests.

CULPHO *(5 m W of Woodbridge)* 'Kull-fo'
Carlford Hundred. AS. *Cuthwulfes hoh* - 'Cuthwulf's promontory' *Culfole* (DB). The Hall stands high upon just such a hill-spur overlooking a tributary of the river Fynn. This is, in all probability, Cuthwulf's ancient site.

DALHAM *(6 m ESE of Newmarket)* 'Dall-um'
Risbridge Hundred. AS. *dael hamm* - 'Homestead in a valley' *Dalham* (DB). Morley suggests the first Hall was in the village down beside the River Kennett and not upon its present site which is high and overlooks the dale - as does the present church which retains no vestige of Saxon work. Both stand upon the junction of the east chalk bank with its crowning boulder-clay, whilst the township itself nestles at the valley's base, formerly with stone bridges over the stream as at Moulton. Originally known as Dalham with Dunstalls, the latter is now the hamlet of Dunstall Green in the SE of the parish.

DALLINGHOO *(4½ m N of Woodbridge)* 'Dah'l'n-hew'
Wilford & Loes Hundreds. AS. *dalinga hoh* - 'The dalemen's promontory' *Dallingahou* (DB). Morley states that 'dalings' were dwellers in a dale. The name may be supposed to have been accorded them when they moved up to the hill-spur represented by the suffixed *hoh*. As one would expect, the Hall here overlooks a stream that rises slightly to the N.

DARMSDEN *(1 m SW of Needham Market)* 'Damm-er-son'
Bosmere & Claydon Hundred. AS. *Deormundes dun* - 'Deormund's hill' *Dermodesduna* (DB). The Hall rises high upon the hillside between church and mere. Once a separate parish, it has from an early period been a hamlet of Barking under the name Barking-cum-Darmsden. It forms a part of the manor of Tarston Hall. The Domesday township of *Bermesdena* - 'Beornmund's Valley' (?) in Claydon Hundred cannot be located but may lie in this area.

DARSHAM *(5½ m NNE of Saxmundham)* 'Dar-sham'
Blything Hundred. AS. *Deores ham* - 'Deor's home' *Dersham* (DB). The literal sense of *deor* is merely a deer. But Morley feels it is here more likely to mean an early nickname of some man who was peculiarly fleet of foot or a noted hunter in the chase.

DEBACH *(4 m NNW of Woodbridge)* 'Debb-idge'
Wilford Hundred. AS. *deopan baece* - 'At the deep valley' *Debenbeis* (DB). The present tiny parish comprises only large flat uplands. Originally it may have been virgin forest and the adjoining parish of Boule (which meets the place-name description), once a part of Debach containing its original church of St Michael on the site of the pre-Conquest church mentioned at Domesday. The present Debach church was built in 1851.

DEBENHAM *(8 m W of Framlingham)* 'Debb-num'
Thedwestry Hundred. AS. *deopan hamme* - 'Settlement by the river between steep banks' *Depbenham* (DB). The town, which suffered severely from a fire in 1744, is now little more than

a large village. In Saxon times, however, it was considerably more important. The Saxon kings of East Anglia occasionally held their courts here, and tradition says that the river Deben was at that time navigable up to the town, despite the fact that it rises only one mile W at Brice's Farm. This would appear to be borne out by its place-name and the fact that an anchor was found embedded in the sand at a place called the Gulls in the last century. The church has evidence of Saxon origins and was one of two recorded here at Domesday. The Domesday *Uluestuna* - 'Wolves farm' is now represented by Ulverston Hall beside the Deben.

DENHAM *(7 m WSW of Bury St Edmunds)* 'Denn-um'
Risbridge Hundred. AS. *dena ham* - 'Homestead in a valley' *Denham* (DB). A valley exists, traversed by a stream running down from Kentford Heath. However, Morley considers it 'Home of the Danes' The original home or hall he thinks is represented by Denham Castle which rises immediately above the 300' contour line; the castle-site (later modified by the Normans) then being the capital stronghold against the Saxons coming from the Cambridgeshire dykes.

DENHAM *(3 m E of Eye)* 'Denn-um'
Hoxne Hundred. AS. *Dena ham* - 'Home of the Danes' *Denham* (DB). The extent of this east Suffolk Denham pretty accurately represent the thirteen hundred acres that constituted 'Hoxne Forest', the possible site where King Edmund was killed by the Danes in 870 A.D.

DENNINGTON *(2½ m N of Framlingham)* 'Dean-y-tun'
Hoxne Hundred. AS. *Denegife tun* - 'Farmstead or village of a woman called Denegifu' *Dingifetuna* (DB). *Denningworth*, anciently a part of Dennington, is named from the AS. *Denegifuworth* 'Denegifu's enclosure' i.e. the actual home of the woman called Denegifu. The site is unknown. The place-name has suffered severe corruption from its original Domesday spelling attaining an *'ington'* for no particular reason but mispronounciation.

DENSTON *(6 m N of Clare)* 'Daan-st'n'
Risbridge Hundred. O.Merc. *Deneheardes tun* - 'Farmstead of a man called Deneheard' *Danerdestuna* (DB). The hall and church lie in the valley of the river Glem.

DEPDEN *(7 m SW of Bury St Edmunds)* 'Depdun'
Risbridge Hundred. AS. *deop denu* - 'Deep valley' *Depdana* (DB). It has been said that there are no hills in Suffolk merely valleys and this is most notable here in the south of the parish where the land dips 120'.

DRINKSTONE *(7½ m SE of Bury St Edmunds)* 'Drinks-'n
Thedwestry Hundred. AS. *drenges tun* - 'Farmstead of a man called Dremic' or 'Soldier's farm' *Drincestuna* (1050); *Drencestuna* (DB); *Drencheston* (1275). Ticehurst House is a possible location for the place-name.

DUNWICH *(8 m NE of Saxmundham)* 'Dunn-idge'
Blything Hundred. AS. *Dunan wic* - 'Duna's village' or 'Deep water' to which *wic* meaning 'harbour' was later added making the probable meaning 'deep water harbour' *Domnoc* (636)?; *Duneuuic* (DB); *Donewic* (1250). Minsmere (locally pronounced 'Miz-ner'), now a part of Dunwich, was the Domesday township of *Milsemere* or *Mensemara*, which Morley suggests could possibly mean 'Lake by the minster'. The lake, doubtless at first inland, would now be salt-marsh after being a haven, following an Act for embanking and draining was made in 1810. The 'minster' possibly referred to the long-drowned see of the Dunwich bishops.

EARL(S') SOHAM *(3½ m W of Framlingham)* 'Arls-soo-um'
Loes Hundred. AS. *saeg hamm* - 'Enclosure by a hollow' *Saham* (DB). 'The church lies low, close by a stream at the base of an abrupt acclivity in a typical Saxon position of shelter. This was the *saeg* or hollow in the land in the same way as we describe a drooping cord 'sags' from the Anglo-Saxon meaning to sink down. After the Conquest, half the village was given by William to Alan, Count of Brittany and husband of his step-daughter Constance. He led the rear at Hastings and became the Earl of Norfolk, Hence in medieval times the township was indifferently termed Saham Comitis, Saham Count or Saham Barres.' (Morley)

EARL STONHAM *(2½ m NE of Needham Market)* 'Arls Stonn-um'
Bosmere & Claydon Hundred. AS. *stan hamm* - 'Homestead by a stone, or with stony ground' *Stanham* (DB). 'Earl' after the Manorial lords, the Earls of Norfolk. The name *stan* meaning 'stony' may be associated with Roman occupation with many finds being made near the church and at Forward Green. The eastern parish boundary follows the A140 Roman road to Caister-by-Norwich.

EAST BERGHOLT *(6 m SE of Hadleigh)* 'East Barflie' *(very early)*, 'Barfold' *(1612)*, 'East Burgolt'
Samford Hundred. O.Merc. *berc*; AS. *holt* - 'Birch copse on or by a hill to the east of another' *Bercolt* (DB). No ancient woodland remains in the parish but the wood referred to may have been east of Kings Wood, Stratford St. Mary. West Bergholt is 'to the west' in Essex.

EASTON *(3½ m S of Framingham)* 'East'n'
Loes Hundred. AS. *east tun* - 'Farmstead or village to the east of another' *Estuna* (DB). Its name no doubt comes from the fact that in early Saxon times Easton was an insignificant village lying to the east of Kettleburgh, then an important settlement. The latter, along with Easton Church, belonged to the powerful Thegn, Edric the Grim of Kettleburgh and constituted the caput of broad lands in many other Suffolk districts. Moated Martley Hall, of which the hall moat remains N of the present village, was a distinct township at Domesday named *Mertlega* or *Martele* - 'marten's meadow'.

EASTON BAVENTS *(N of Southwold)* 'Est'n Bev'n'
Blything Hundred. AS. *east tun* - 'Farmstead to the east of another' *Estuna* (DB). Now lost to the sea but formerly the most easterly point of the British Isles and termed *Exoche* by the Romans. Hubert de Bavent was lord here in 1263 and Thomas Bavent the richest man in 1327. Their name was added to Easton in 1523 no doubt to distinguish the parish from that near Framlingham.

EDWARDSTONE *(5 m E of Sudbury)* 'Edd-es-t'n'
Babergh Hundred. O.Merc. *Eadwardes tun* - 'Edward's farm' *Eduardestuna* (DB). The place can claim no history prior to the Normans and the church retains no Saxon architecture. Edwardstone Hall, taken down in 1952, probably occuped the site of the place-name and Domesday hall of the wealthy De Montechensy family. The name's final *-e* is recent and superfluous.

ELLOUGH (or Willingham All Saints) (see also Willingham St Mary) *(3½ m SE of Beccles)* 'Ell-ler'
Wangford Hundred. O.Norse. *Helga* - 'Helgi's place' or 'Place at the heathen temple' *Elgar* (DB). *Elgr* is the Norse word for temple. The church of All Saints may have replaced a heathen one - possibly on the same site(?)

ELMSETT *(4 m NNE of Hadleigh)* 'Emm-set' or 'Ull-um-set'
Cosford Hundred. AS. *elme saetan* - 'Dwellers among the elm-trees' *Elmeseta* (DB). Just where the original elm trees may have stood cannot be ascertained now, but the site of the Hall near Belstead Brook running down from Naughton is their most likely location.

ELMSWELL *(5 m NW of Stowmarket)* 'Emms-wool'
Blackbourn Hundred. AS. *Aelfmaeres wella* - 'Aelfmaer's well' *Elmeswella* (DB). Elmer is a known surname in the area. A spring was shown at Hawks End, S of the Hall and site of the former grange of the Bury Abbots, on earlier OS maps. The village may even, at that time, have encompassed the celebrated Lady's Well which lies SW, (now) 400 yds. within Woolpit parish.

ELVEDEN *(5 m S of Brandon)* 'Elden' or 'Ell-d'n'
Lackford Hundred. AS. *Aelfan denu* - 'Aelfa's hill or valley' *Eluedena, Heluedana, Haluedona* (DB). The various Domesday spellings mean that the word could be *-den*, a valley, or *-don*, a hill. Amazingly, a fishery is listed here in 1086, no doubt on the western parish boundary where the Eriswell Lodge Drain now drains the fenland which anciently formed the Old Fen Sea.

ERISWELL *(3 m N of Mildenhall)* 'A-zell' now 'Err-ez-well'
Lackford Hundred. O.Merc. *Eferes wella* - 'Efer's well' *Hereswella* (DB). At Eriswell Lodge a brook still bears the Saxon name of the Harst (*haest* - 'violent'). Two separate Pagan Saxon settlements developed here beside two wells or streams. Caudle, the Domesday *Coclesworda*, formerly a berewick (secondary manor), became Cocilsworth with its own hall and church with, south of this, Eriswell, also with its own church. Already in severe decline with its hall and church in ruins, Cocilsworth was eventually obliteratd by the construction of the Lakenheath airbase, leaving a small hamlet now known as Little Eriswell. The village of Eriswell, near the western boundary, was then left to oversee this massive parish of fen edge and Breckland heath and fir.

ERWARTON (formerly Arwarton) *(9 m SE of Ipswich)* 'Are-wutt-'n'
Samford Hundred. O.Merc. *Eforwardes tun* - 'Eforward's farmstead' *Eurewardestuna* (DB). Morley also places the lost Domesday *Eduinestuna* - 'Eadwine's farmstead' here.

EUSTON *(4 m SE of Thetford)* 'Evston' (*early*), 'U-stun'
Blackbourn Hundred. AS. *Eofes tun* - 'Eof's Farmstead' *Euestuna* (DB). Little Fakenham - *Litla Fachenham* (DB); *Fakenham Parva* (1340), is a hamlet consolidated with Euston in 1669. Rushford - *Riseurda* (DB), partly in Norfolk, is another Euston hamlet. All its old spellings show Rushworth - 'the rushy tenement', but this was probably changed when the Little Ouse River attenuated to a fordable depth.

EXNING *(2 m NW of Newmarket)*
Lackford Hundred. AS. *Gixaninga* - 'Place of Gixa's family' *Essellinge* (DB); *Exningis* (1158). Before the draining of the fens it was, as Morley puts it, 'our westward land's end with 30 miles of morass across to St. Ives and 10 to Ely. He also notes that the Domesday spelling also suggests the AS. *essel* 'a shoulder' with the diminutive *-ling*; and Exning certainly pushes a little shoulder out westward into the level fenland.

EYE *(5 m SSE of Diss)* 'Oy' as in "boy" or (earlier) 'Ay' - *the Suffolk name of d'Eye always pronounced 'Day'.*
Hartismere Hundred. Angl. *eg*; AS. *ieg* or *ig*. 'Dry ground (island) in the marsh' *Eia* (DB). As its name suggests, the town was once a virtual island, described in the latter part of the C10 by Abbo

Floriacencis as situated in the midst of a marsh and navigable from the Waveney. This was confirmed by finds of boating artefacts in the surrounding fields. In pre-Conquest times the town obviously lay amid one of the largest sheets of water in the whole county (the southern low bogland is still locally termed The More i.e. Mere, and the particular one which, almost certainly, gave name to Hartismere Hundred). Further S on high ground still stands the Hall of Cranley, the Domesday manor of *Cranlea* - 'Crane's meadow'. *Suddon* - 'south hill' was listed as a hamlet of Eye in 1327.

EYKE *(3½ m NE of Woodbridge)* 'Ike'
Loes Hundred. O.Norse. *eik* - 'Place at the oak-tree' *Eik* (1185). No apparent reason emerges for the absence of both Eyke and Iken from Domesday Book; both names, if not places, date from before the Conquest. As a place-name Eyke "was not in existence" in 1086 (Suff Inst. 1898, 71); and it probably arose when "a Norman church was built about 1150 in the village of Eyke, which had only just come into existence". However, at that time it may well have been represented by Staverton, now a mere hamlet of Butley. Staverton Forest is situated here and in the adjoining parish of Wantisden. It was from this ancient woodland, once much more extensive, that Eyke attained its Scandinavian place-name.

FAKENHAM MAGNA *(5½ m SE of Thetford)* 'Fake-num'
Blackbourn Hundred. AS. *Facan hamm* - 'Facca's enclosure' *Fachenham* (DB) *Fakenham Magna* (1283). Facca's enclosure may refer to the earthworks known as Burnt Hall Plantation which lies in a meadow on the N bank of the river Blackbourne ¼ mile S of the church. The Saxon Manor of Fakenham was held by Alestan and was given by William I to his nephew Peter de Valoines after the Conquest. The field names and Saxon place-name are suggestive of Facca's pre-Conquest defensive 'enclosure' probably later utilised by the Normans as a ring-works (a form of early castle) and perhaps occupied by the Valoines' who held the manor for some time.

FALKENHAM *(9 m SE of Ipswich)* 'Fork-num'
Colneis Hundred. AS. *fealcan hamm* - 'Falcon's enclosure' *Faltenham* (DB). Falkenham, as part of the ancient port of Gosford, was a more important district in 1086 than now, and its church's dedication to St. Aethelbeorht seems to place its original erection to *c*793 AD.

FARNHAM *(3 m SW of Saxmundham)* 'Farn-um'
Plomesgate Hundred. AS. *fearn hamm* - 'Homestead or enclosure where ferns grow' *Ferneham* (DB). Whilst the parish now comprises little more than the church, hall, rectory and a cluster of properties on the northern boundary, the original settlement may well have centred around the river Alde crossing at Langham Bridge. Here an ancient island site known as Farnham Bottley - a corruption of 'Boat island' - may have housed the first 'enclosure', whilst a C13-C15 pottery scatter points to continued medieval settlement in this area.

FELIXSTOWE 'Feel-ix-tow'
'Colneis Hundred. AS. *fylleth stow* - 'A place of felled tress or a clearing' (Skeat) *Filchestou* (1254); *Filthstowe* (1316). Though unlikely, many are of the opinion that St. Felix landed and set up his see in the now lost Roman fort at Walton and that the place-name comes from this fact. The Manor of Walton, held by Roger Bigod, once extended over the greater part of the present Felixstowe parish forming the main peninsular limb of the important port of Gosford. Felixstow was a mere village up to the mid-C19, and the adding of the final *e* in later spelling is quite shown

meaningless. Langer, a clipt form of Langerston, the AS. *lang stane* - 'long stone' *Langestuna* (DB), was a 2-mile sand-ridge protecting the Orwell estuary. *Poleshead*, the AS. *Pol's heafod*, was a now-lost island off Landguard Point. Domesday Book also shows *Burgesgata* and *Mycelegata* to have been distinct estates here probably both protecting approaches ('gates') to Walton Castle.

FELSHAM *(8 m SE of Bury St Edmunds)* 'Fell-sam or 'Fill-shum'
Thedwestry Hundred. AS. *Faeles ham* - 'Faele's home' *(Faele* 'good or faithful') *Fealsham* (DB). Faelew's home may have occupied the site of the present Felsham Hall which stands within an ancient moat, but Capel Farm, 1¼ miles SW on the parish boundary, is built on the site of an earlier C13 Felsham Hall.

FINNINGHAM *(7½ m N of Stowmarket)* 'Finny-gam'
Hartismere Hundred. AS. *feningham* - 'Fenmen's home' *Finingaham* (DB); *(Feningham* (IPM); *Fynigham* (1327 Subsidy). This cannot be 'The homestead belonging to Finna's people' as so often stated, as the *Finningas* 'sons of Finn' had double *nn*, whereas the Domesday Book and all early spellings were with single *n*. The name more likely began as *fening*, 'the men of or from the fen'. The spot, now marked by Sto(ke)land or Sto(ne)land "Abbey" (where there was never a monastery), seems to have been as isolated as was Rymer in Fakenham at the time of parishes demarcation; for there now five townships meet on the west Suffolk border, and here five eastern ones meet at Allwood Green (formerly Aldwood Forest). It is interesting that both sites are marked by ring earthworks; Burnthall Plantation in Fakenham and Cromwell's Plantation in Finningham. Felmingham, the Domesday *Felincham* belongs here. It seems associated with *felnys* - 'fierceness.'

FLEMPTON *(5 m NW of Bury St Edmunds)* 'Flam-t'n' now 'Flem-t'n'
Thingoe Hundred. O.Merc. *fleminga tun* - 'Farmstead belonging to the fugitives' *Flemingtuna* (DB). The intrusive and wrongly added *p* to Flemton dates from both the Nonarum of 1340 and Poll Tax of 1381 but the pronounciation has remained true.

FLIXTON *(2½ m SW of Bungay)* 'Flick-st'n'
Wangford Hundred. Dan. *Flikkes*; AS. *tun* - 'Flik's farm' *Flixtuna* (DB). Flik's farm may well have occupied the moated site of Boys Hall, S of the church, which is thought to have preceded Flixton Hall which stood close by.

FLIXTON *(3 m NW of Lowestoft)* 'Flick-st'n'
Mutford & Lothingland Hundred. Dan. *Flikkes*; AS. *tun* - 'Flik's farm' *Flixtuna* (DB). The site of the old Hall, that doubtless of FitzOsbert's main Manor of Flixton, stands near the apex of a hill - a likely Danish site - with the ruined church immediately SW.

FLOWTON *(6 m NW of Ipswich)* 'Flo-t'n'
Bosmere & Claydon Hundred. AS. *flocc tun* - 'Flock (sheep) farm' *Flochetuna* (DB). Flowton Hall is the likely site for the former sheep farm situated above the brook in this tiny parish.

FORNHAM (ALL SAINTS, ST MARTIN, ST GENEVIEVE) *(N & NW of Bury St Edmunds)* 'Forn-um S'int' etc.
Thingoe Hundred. AS. *Fornan ham* - 'Forna's home' plus church dedication *Fornham* (DB). Now comprises three parishes. Alnothus, an officer of Edward the Confessor, is said to have bestowed All Saints; and King Edmund (940-946 AD), son of Edward the Elder, St. Martin and St. Genevieve. All Saints includes Babwell (at first *Badwell* - 'Bada's well or pool') with its medieval mill and friary - "To this manor belonged a mill and parts of Babwell Fen". Aerial photographs have

what appears to be a cursus 'Ceremonial Way' running parallel with the A1101 road from S of All Saints Church for a mile to Hengrave which may suggest an ancient pre-historic function with a later ecclesiastical union of the Fornhams.

FOXHALL *(2 m SE of Ipswich)* 'Fox-ul'
Carlford Hundred. AS. *foxa holu* - 'Holes of foxes' *Foxehola* (DB). Foxhall Hall here is, consequently, not taurological but is a mispronounciation of Foxhole Hall. The parish possibly included the former Domesday vills of *Ingoluestuna* - 'Ingwulf's Farm', *Derneford* - 'Danes' Ford' (a footbridge crossing of this river Deben tributary leads N to Pole Hill bowl barrow), *Isleuestuna* - 'Iseleof's Farm', and *Necchemara* - 'Nectan's Mere' (possibly now Bixley Decoy).

FRAMLINGHAM 'Fram-e-gun' and 'Frann-ing-um'
Loes Hundred. AS. *Framlinga ham* - 'Home of Fram's family' *Framelingaham* (DB). Tradition would suggest that Framlingham Castle was built on an existing mound which formerly held a Saxon stronghold (of Fram's family?) but there is no evidence for this and a cemetary discovered in the outer bailey in 1954 and thought Saxon was more probably medieval.

FRAMSDEN *(11 m NE of Ipswich)* 'Frams-d'n'
Thedwestry Hundred. AS. *Frames denu* - 'Fram's valley' *Framesdena* (DB). The original site may improbably be marked by the present Hall between the river Deben and one of its tributaries by Jockey ("Jockey of Norfolk," i.e. Duke Howard of Bosworth Field?) Lane here; but not in the valley, which is a deep one. Upon a conspicuouse hoh, jutting into an angle of the tributary, is a more likely settlement for the earlier Hall.

FRECKENHAM *(4 m SW of Mildenhall)* 'Freck-num' and 'Freck-en-um'
Lackford Hundred. AS. *frecan hamm* - 'Homestead or village of a man called Freca' or 'Warriors' enclosure' *Frekeham* (895); *Frakenaham* (DB). Morley's suggested that the place-name refers to 'the enclosure of a band of warriors', a reference, perhaps, also to the earthworks in Mount Meadow (converted later into the Norman motte and bailey castle). There were also remains of an isolated dry rectangular moat (now ploughed out) in the suggestive Mot Meadow to the E of Mortimers Lane which runs N from the village. He argues that 'the name is no less suggestive than the situation and physical features of this place, on the very edge of the erstwhile waters of the Old Fen Sea which covered a large extent of NW Suffolk in ancient times. This is the most westerly point of the county that was land a thousand years ago. Three miles away King Anna's daughter Aethelthryth was born at Exning in 630; four miles away Bishop Felix was buried at Soham in 647; and, on the edge of the Breck, some ten miles westward is Ampton, which may have been Anna's town. When Penda invaded Suffolk in 654 he possibly sailed from Mercia cross the Fen Sea, upon whose shores here data is accumulating of the East Angles' defence. Whence I consider we are justified in regarding the *ham* as meaning enclosure and not home, for obviously still the river Kennet here has been artificially diverted.'

FRESSINGFIELD *(9 m W of Halesworth)* 'Frez'nfield'
Hoxne Hundred. AS. *Fresena feld* - 'Frisian's clearing' *Fessefelda* (DB). The place has carried the incorrect inclusion of -*ing*- from 1275 to the present day. It should be Fressenfield. Domesday shows two considerable hamlets here. The first variously spelt *Cebbenhala; Cibbenhala; Cybenhalla* and *Cipbenhala* is now Chippenhall - 'Hall or shelter of Ceapa'. The second, *Wettingham*, now Whittingham - 'Home of Wiht's family'. The Domesday *Icheburna* - 'Thick

broom-plants?' also relates to this parish forming part of the Manor of Veales with Launes and Thykbrome Veales cum membris, the lordship of William de Veel in 1275.

FRESTON *(3½ m S of Ipswich)* 'Free-t'n'
Samford Hundred. AS. *Fresan tun* - 'The Frisians' farm' *Frisetuna* (DB). The positions of Freston, Friston and Fressingfield rather tend to argue of a late date for settlement; the very narrow foreshore of the present place looks as though it had been annexed from Woolverstone.

FRISTON *(2¼ m SE of Saxmundham)* 'Friss-t'n'
Plomesgate Hundred. AS. *Fresan tun* - 'The Frisians' farm' *Frestone* (1327). Morley feels that there is something inexplicable in the present name, which does not occur in Plomesgate Hundred at all throughout Domesday Book. On the other hand there were in 1086 two estates in this Hundred called *Bohtuna*, perhaps meaning a 'branch' farm or Berewick possibly of Snape; and one estate called *Riscemara*, the modern Rushmere; neither developed into a medieval manor. For lack of a better position Morley guesses them to represent Friston. The name *Polsborough Gate*, recorded by Kirby in 1735 for the junction of the four cross roads just N of Friston Decoy, must have applied to a conspicuous tumulus (*'Pfol's* barrow') which the plough has swept away and of which modern maps do not give even the name.

FRITTON *(8 m NE of Beccles)* 'Fritt-'n'
Mutford & Lothingland Hundred. AS. *frith tun* - 'Farmstead offering safety or protection'. A town wherein *frith*, or peace was secured between two contending parties was termed *frith-burg* by the Saxons; there is a Frith Wood in Cowlinge and here we have the farm of peace or *frith tun*. The security may have been due to situation for none better could be found than snugly berthed in Fritton Decoy on the landward side of Lothingland penisula. Though any define basis for it seems lacking, we must not forget that 'tradition tells us that the name of Fritton Lake was once *Gunhelda's Mere*. Gunhelda was sister to Sweyne, the great Danish pirate; and it is said she was put to death on 13 Nov 1002 ('beheaded by the express direction of Edric Streone' rather then renounce her religion). It is also thought that Sweyn's son, King Canute (Knut), founded the church - the round tower is very ancient. Two minor hills have been left on the summit of the large Bell Hill mound as a result of digging into what was thought to be a burial mound. However other suggestions have been voiced as to the origin of this large hill or mound. One is that it is in fact a Danish earthwork created by cutting through a natural ridge to create a defensive post some 30 paces from reclaimed cattle marshes. The name has been taken from the adjoining village of Belton (*Beletuna* at Domesday). *Beli*, a Viking, it is further speculated, first took this fort by assault and subsequently settled in the township which now bears his name. It was noted that there was a large percentage of Norse place-names in Lothingland Hundred. Caldecot Hall represents the Domesday Manor of *Caldecotan* AS. *ceald cote* - 'Cold and sheltered dwelling' comprising one carucate (120 acres).

FROSTENDEN *(4½ m NW of Southwold)* 'Froz-en-d'n', *pronounced as at Domesday*
Blything Hundred. AS. *froscan denu* - 'Valley frequented by frogs' *Froxedena* (DB). The intrusive *t* was added from 1250. Now under the plough but clearly visible, a circular mound (just across the parish boundary in South Cove) overlooks to the W what remains of a possible ancient dock and quay dug out of the E side of the river entering at Easton Broad. Domesday designated *Froxedena* a sea-port (*Portus Maris*), and Suckling suggests that this was in the area of Frostenden Bottom through which a small stream winds its way by South Cove to Easton Broad,

34

tracing the parish boundary with South Cove. The river was much wider and deeper in earlier times. It may have Danish origins but its purpose and that of the adjacent mound have yet to be confidently established. Domesday records a saltworks in the parish, probably at Frostenden Bottom. This was discontinued soon after, suggesting that the sea-port was already well in decline by this time.

GAZELEY *(5 m SE of Newmarket)* 'Gaz-ly' *as in* 'gay'
Risbridge Hundred. AS. *Gaeges leah* - 'Gaeg's woodland clearing' *Gaysle* (1275). Gazeley was not a parish at Domesday but the manor of Desning covering 4,400 acres with berewicks in Cavenham, Lakenheath and Mildenhall, and probably included (though not among its rateable acres, of course) the erstwhile extraparochial Southwood Park, now in Hargrave. The modern Desning Hall (1845) SE of the village, stands beside the great moated site of the ancient manor house of 'Deselynge', the probable original settlement site.

GEDDING *(8 m W of Stowmarke* 'Giddin' *(old)*
Thedwestry Hundred. AS. *Gyddinga* - 'The place belonging to Gydda's people' *Gedinga* (DB). Gedding Hall lies isolated N of the church in a defensive position on the summit of rising ground. In such a small parish this would seem the likely site of Gydda's people.

GEDGRAVE *(½ m SSW of Orford)* 'Gigg-rivv' or 'Giggeruv'
Plomesgate Hundred. AS. *geat graef* - Obscure in origin. It may mean 'Goat's enclosure' but Morley suggests 'Ditch of passage' *Gategraua* (DB). He suggests that the fairway between the shore and Havergate Island, which passage (The Gull) has altered only in detail since at least the drawings of Norden's 1601 map, is exactly such a ditch of passage. The spelling *Gategrave* is fairly uniform from 1086 to (Orford town deeds) 1388.

GIPPING *(4 m NE of Stowmarket)* 'Gipp-en'
Stow Hundred. AS. *Geppinga* - 'The place of Geppa's people' *Gippinges* (C12). This village gave name to the river Gipping and not conversely, for whereas a village can belong to a family, a river cannot, and the Saxon *-ing* means 'the family of'. Nor does the river actually rise here but in the neighbouring hamlet of Mendlesham Green.

GISLEHAM *(5 m SW of Lowestoft)* 'Giz-el-um'
Mutford & Lothingland Hundred. Norse. *Gisle*; AS. *ham* - 'Gisli's homestead' *Gisleham* (DB). S of the church, the site of the Manor House still retains an impressive double moat, one inside the other. The manor formerly belonged to the knightly Garney family.

GISLINGHAM *(5 m SW of Eye)* 'Gizel-gam' now 'Giz-lenn-um'
Hartismere Hundred. Norse. *Gisl*; AS. *inga ham* - 'The home of Gisli's family' *Gyselingham* (1060); *Gislingaham* (DB). Isaac Taylor refers to a theory that the leader and elder of the family would arrive and settle near the coast (as at Gisleham) and that his sons (except the younger who, in those days, succeeded) would later leave to create a new settlement inland (as at Gislingham).

GLEMSFORD *(5 m NE of Clare)* 'Glemms-fud'
Babergh Hundred. ?AS. *claemes ford* - The meaning is obscure and encourages several translations. Skeat thinks it 'ford over the gleaming water' (though there is no such Saxon word as *glaem* to be known). Morley goes for 'Clam's ford' - these large and conspicuous molluscs being abundant in stagnant water in many Suffolk localities; whilst Mills translates it as 'ford where people assemble

for games and revelry' *Glemesford* (c1050); *Clamesforda* (DB). Could the site be where Scotchford Bridge now carries the B1065 over the river Glem, here little more than a stream? Two roads and a footpath from Stanstead Church meet here at the river.

GOSBECK *(7½ m N of Ipswich)* 'Gorz-brook' or 'Gorz-beck'
Bosmere & Claydon Hundred. AS. *gos baec* - 'Stream frequented by geese' *Estuna* (DB); *Gosebech* (1179). Our *baec* here is represented by two minor affluents that merge in the river Gipping by Bosmere Hall. The parish was originally called Easton. To distinguish it from two others of that name in Suffolk it assumed the name Easton Gosbeck in 1174. In the C17 it finally dropped the name Easton altogether to become plain Gosbeck - hence its ommission under this name at Domesday, where it is recorded with 14 acres as Easton in Bosmere Hundred. The Domesday *Langedana; Langhedana,* and *Langhedena* (AS. *lang denu* - 'Long valley') has been synonymised with 'Langham in Gosbeck', presumably a gully running down to Stonewall Farm.

GREAT ASHFIELD *(8 m NW of Stowmarket)* 'Ash-feld'
Blackbourn Hundred. AS. *aesce feld (magna)* - 'Clearing where ash trees are felled' *Eascefelda* (DB). The addition of Great (*Magna*) was not required and hardly ever used as Little (*Parva*) Ashfield became popularly known as Badwell Ash - though there is an Ashfield (Ashfield-cum-Thorpe) in Thredling Hundred.

GREAT BARTON *(2½ m NE of Bury St Edmunds)* 'Bar-t'n'
Thedwestry Hundred. AS. *bere tun (magna)* - 'Large corn or barley farm' *Bertuna* (DB). Magna (*Great*) was added to distinguish it from Little (*Parva*) Barton, though this, too, is superfluous as the latter is now known as Barton Mills and located in Lackford Hundred. Includes the Domesday estate of *Cattishall*, perhaps another of those given by Earl Ulfketel to Bury Abbey.

GREAT BEALINGS *(2½ m W of Woodbridge)* 'Graat Ballens or Beelens'
Carlford Hundred. AS. *Beolingas (magna)* - 'Settlement of the Beola (tribe)' or 'Settlement by the funeral pyre' *Belinges* (DB). In the park NE of Rosary Farm and S of Bealings House, 'an ancient mound which has yielded up pottery, two urns and flint tools, and is supposed to have been British, has been found in the parish'. This may suggest an alternative place-name of 'Settlement of the dwellers by the funeral pyre.' Seckford is situated in this parish, AS. *Soec-forda* - 'The warrier's ford' or AS. *Secg-forda* - 'The sedge ford' *Sekeforda* (DB).

GREAT BLAKENHAM *(5 m NW of Ipswich)* 'Graat Blake-num'
Bosmere & Claydon Hundred. AS. *Blacan ham (magna)* - 'Blaca's homestead or enclosure' *Blac(he)ham* (DB). Now divided into Great and Little Blakenham but at Domesday it appears to have contained but a single township and not until the C14 does its bisection emerge. It is therefore not possible to site its five Domesday entries.

GREAT BRADLEY *(6 m N of Haverhill)* 'Bradd-ly'
Risbridge Hundred. AS. *braden leah (magna)* - 'Broad meadow' *Bradeleia* (DB).

GREAT BRICETT *(7 m S of Stowmarket)* 'Bri-set' *with long 'i' as in 'ice'*
Bosmere & Claydon Hundred. AS. *beorht saete (magna)* - 'Settlers on a bright spot' or 'Fold or stable infected with gadflies' *Brieseta* (DB).

GREAT CORNARD *(1 m SE of Sudbury)* 'Korn-ud'
Babergh Hundred. AS. *corn earth* - 'Cultivated land used for growing corn' *Cornerda* (DB). Cornard was anciently called 'Corn-earth' and became the property of a family who assumed the

name. John de Corn-earth was Sheriff of Norfolk and Suffolk from 1206 to 1209. The AS *eorthe* means 'earth' but without the final *e* its meaning becomes 'plough-land'. Great and Little Cornard were not divided at Domesday though two manorial churches are given and it is now impossible to allot the five parcels of land, whereof one had 303 cultivated out of 320 acres with woodland, and another had no more than 126 cultivated out of a total of 960 acres with woodland.

GREAT FINBOROUGH *(2½ m SW of Stowmarket)* 'Finn-brah' or 'Fimm-brah'
Stow Hundred. AS. *fin beorh (magna)* - 'Heaped-up mound' or 'Fin's tumulus' *Fineberga* (DB). The hill or mound of the place-name is Devil's Hill which is situated on a golf course in Finborough Park in the NW corner of the parish. Tradition states that a part is artificial, raised as a defence in Saxon times. The 'Battle of Finnesburn' is said to have been fought out here when Danes falling back from Thetford in 1005 encountered the Saxon army. "For five long days the battle's sound was heard by Finburghe's earth-raised mound." Eventually, the determined force of defenders won the day. The wounded leader of the attackers withdrew his battle-stained forces declaring that "they had never met a worse hand-play" and left behind a tiny but victorious garrison, weary and wounded, but with spirits "undiminished and unquell'd." Unfortunately, tradition is all we have to go on as there is no evidence for the event and its site is also claimed by Playford on the river Finn. Borough ('a town') has been erroneously substituted for *bergh*, 'a hill'.

GREAT GLEMHAM or Glemham Magna or North Glemham *(4 m W of Saxmundham)* 'Graat Glemm-un'
Plomesgate Hundred. ?AS. *Glaem hamm* - The meaning is obscure and several authorities have had their say without real conviction: Skeat favours 'Gleam-enclosure' (from the AS. *glaem* - 'brightness' i.e. 'a sunny situation' comparing our other village Glemsford - 'Ford through the Gleam River'). Reyce's Breviary of 1618 claims that the river Alde was earlier termed the Gleme 'which cometh from Rendlesham (Rendham?) and both the Glemhams' *Glaimham* (DB).

GREAT LIVERMERE or Livermere Magna *(5 m NNE of Bury.)* 'Livv-mere'
Thedwestry Hundred. AS. *laefer mere* - 'Yellow Iris mere' *Liuermera* (DB). The lake becomes obvious when we notice that there a tributary of the river Lark rises in a broad splay and forming an ornamental mere in Livermere Park. The prefix is a misnomer as the parishes are similar size; Great Livermere occupies 1500 acres and Little Livermere, 1400. They are also in different Hundreds with the line of demarkation running down the centre of the lake emphasising its early importance.

GREAT SAXHAM *(5 m SW of Bury St Edmunds)* 'Gritt Saxum'
Thingoe Hundred. O.Merc. *Saxan ham* - 'Seaxa's homestead' *Saxham Magna* (DB).

GREAT THURLOW or Thurlow Magna *(4 m NE of Haverhill)* 'Grate Thurler'
Risbridge Hundred. AS. *Thrythe hlaw* - 'Lady Thryth's Tumulus' *Tridlauua* (DB). Both its place-name and ancient spelling 'Tritlaw' - meaning famous tumulus or assembly hill, indicate a large burial mound situated in the parish, probably later utilised as a meeting place, and at present unlocated. It may have been used as an administrative centre for the rural community of what became in 1044 the 'Liberty of St Edmund'.

GREAT WALDINGFIELD or Waldingfield Magna *(3 m NE of Sudbury)* 'Wonnerfeld' or 'Wannerful'
Babergh Hundred. O.Merc. *Waldingafeld* - 'Walda's family clearing' *Wealdingafeld* (c995); *Waldingefelda* (DB).

GREAT WENHAM or Wenham Magna *(4½ m SE of Hadleigh)* 'Wenn-um'
Samford Hundred. AS. *Wenan ham* - 'Wena's Homestead' *Wenham* (DB). A single township until after the Conquest, it was first divided in the 1327 subsidy, when it was called *Wenham Combust*, and later often *Brent Wenham*. In the C18 Kirby calls the parish *Wenham Magna* and *Burnt Wenham*. Vauxhall, 1½ miles N of the village, site of the Vaux's original manor house and named Farce Hall by Faden in 1783, may stand within Wena's ancient circular moat.

GREAT WHELNETHAM or Welnetham Magna *(4 m SSE f Bury St Edmunds)* 'Weltham' or 'Grate Well Neeth-um'
Thedwestry Hundred. ?AS. *hweol Witan ham* - 'Wita's circular home?' or 'Enclosure frequented by swans near a water-wheel (or some other circular feature)' *Hvelfiham* (DB); *Weluetham* (1170). One township at Domesday, the place-name may refer to the ancient mere in the hamlet of Sicklesmere, N of the parish (and/or a possible ancient watermill).

GREAT WRATTING or Wratting Magna *(2½ m NE of Haverhill)* 'Graat Ratt'n'
Risbridge Hundred. AS. *Wraettinga* - 'Of Wraett's family' *Wratinga* (DB). Along with Little Wratting (*Wratting Parva*) and West Wratting (in adjacent Cambridgeshire), Wratting formed one large single estate in early Saxon times.

GROTON *(6 m W of Hadleigh)* 'Graw-t'n' or 'Grow-t'n'
Babergh Hundred. ?AS. *grotan* - 'Sandy or pebbly stream' or 'place where oats are hulled (coarsely ground)' *Grotena* (DB). A tributary of the river Box flows northwards through the parish.

GRUNDISBURGH *(3½ m NW of Woodbridge)* 'Grund-es-borough'
Carlford Hundred. O.Norse. *Grundes*; AS. *burh* - 'Grundi's stronghold' *Grundesburch* (DB). 'The Stronghold' - an Iron-Age Camp overlayed by an early Roman settlement - is in the adjoining parish of Burgh, once a part of Grundisburgh, i.e. named by the Saxons *Grundi's - burgh*.

GUNTON *(1½ m N of Lowestoft)* 'Gun-t'n'
Mutford & Lothingland Hundred. O.Norse. *Gunna tun* - 'Gunni's farmstead' *Guneton* (1198). Not recorded in Domesday but *Duneston* 'the farm by the coast-sand-hills' which has two entries and is in the 1275 Hundred Rolls is identified as being a hamlet of Gunton-next-Oulton. In 1086 it had only 60 rateable acres with woodland for but four swine. No record of its obvious submergence or reason for Gunton's omission are discoverable.

HACHESTON *(4 m SE of Framlingham)* 'Ha-chess-t'n'
Loes Hundred. AS. *Haeces tun* - 'Haec's farmstead' *Hacestuna* (DB). Includes the Domesday vill of *Glereuinges* (now Glevering) from where the once powerful family of De Clavering took their name in 1313. Probably from the AS. *glafan* 'to glow' in respect of the shining waters of the Deben here, or perhaps from the Scots word clavering 'gossip'. Locally pronounced 'Glav-r-en'.

HADLEIGH 'Hadd-ley'
Cosford Hundred. AS. *haeth leah* - 'Heath meadow' *Hetlega* (DB). Hadleigh Heath was enclosed in 1832. The settlement of Hadleigh was created in a 'clearing in the heath', obviously referring to this stretch of open land which must at that time have been very extensive. Fadens 1783 map shows it considerably reduced and on enclosure only comprising some 20 acres. The parish includes *Stanestrada* - Stone Street, *Topesfelda* - Topsfield, and *Latham* - Layham-in-Topsfield, wherein lay only a 20-acre berewick of Churchford manor in Capel. Layham means 'sheltered enclosure (in the Field of Tope)'.

HALESWORTH 'Holzer' *(early), now* 'Hails-worth'
Blything Hundred. AS. *Haeles worth* - 'Hael's enclosure' *Healesuurda* (DB). Situated on the Roman Stone Street to Bungay and no doubt a place of some importance in early Saxon times.

HARGRAVE *(6 m SW of Bury St Edmunds)* 'Har-grov' or 'Har-gruve'
Thingoe Hundred. AS. *haran grafa* - 'Hare's/Harold's grave or ditch' *Haragraua* (DB). Means literally 'hare's grave' but more likely than that of an animal is the grave of a man nicknamed Hare or perhaps the tumulus of a Danish Viking called Harold? The site is unknown.

HARKSTEAD *(7 m SE of Ipswich)* 'Harrk-sted'
Samford Hundred. O.Norse. *Haerekr*; AS. *stede* - 'Herekr's place' *Herchesteda* (DB). The Hall here (unlike Saxon ones in sheltered valleys) is pretty surely upon its Norse site, occupying a distinctly strategic position now 1¼ miles inland upon high ground overlooking the river Stour at both Beaumont Hall and Holbrook Bay. To the former inlet a rivulet runs down from Redhouse Farm, diagonally across the front of Harkstead Hall, whereto it and its valley afforded additional defence in early times.

HARLESTON *(3 m NW of Stowmarket)* 'Harl-st-t'n'
Stow Hundred. AS. *Heoruwulfes tun* - 'Heorwulf's farmstead' *Heroluestuna* (DB). The hall stands close to the little pre-Norman church dedicated to St. Augustine.

HARTEST *(8 m NW of Sudbury)*
Babergh Hundred. AS. *hert hyrst* - 'Wooded hill frequented by harts or stags' *Herterst* (DB); *Herthurst* (1496). Ashen Wood and Bavins Wood on the eastern parish boundary form part of ancient woodland. Overlooking the pretty village sitting snugly in the valley, they would appear to readily fit that described in the place-name, although with deer so plentiful in the woods of England in Saxon times Morley hypothasises that some peculiar incident of their chase must have taken place here to render the present area unusually appropriate to the title.

HASKETON *(1 m NW of Woodbridge)* 'Hask-e-t'n'
Carlford Hundred. Norse. *Hoskuldr*; AS. *tun* - 'Hoskuldr's farmstead' *Haschetuna* (DB). In this village lie two other Domesday places of Carlford Hundred. *Torp*, the later manor of Thorp Hall, consisting of three carucates (360 acres) and *Hopestuna* or *Hobbestuna* - perhaps 'the farm of Hop or Hobb' but more likely AS *hop*, meaning a hoop, which makes Hopeston or 'Farm by the Osiers' of which hoops were constructed, doubtless in a marshy site by the Hasketon tributary of the river Lark.

HAUGHLEY *(3 m NNW of Stowmarket)* 'Hor-ly'
Stow Hundred. AS. *hagan leah* - 'Enclosed meadow' *Hagele* (c1040); *Hagala* (DB). In 1065 *Hagala* comprised a single manor of 1,120 acres with enough forest for 200 swine, among the largest in Suffolk and belonging to Thegn Guthmond of Stanstead, brother of Abbot Wulfric of Ely.

HAVERHILL 'Hav-rill' or 'Have-rill'
Risbridge & Hinckford (Essex) Hundreds. AS. *haefer hylle* - 'Billy-goat hill' *Hauerhella* (DB). Most claim it to be Scandinavian and mean 'Hill where oats are grown', but Morley states that *haver* meaning oats was unknown before 1300. The place-name must refer to the W of the parish which stands high (for Suffolk).

HAWKEDON *(9 m SSW of Bury St Edmunds)* 'Hork-e-dun', also 'Harden'
Risbridge Hundred. O.Merc. *Hafoc run* - 'Hawk's hill' *Auokeduna; Hauccbenduna; Hauokeduna* (DB). The inconsistent spelling of this township in Domesday leads Morley to the conclusion that 'Hawk' was a personal name. Hawkdon Hall itself lies very low beside the river Glem and so undoubtably on its original site that the place was at first pretty surely Hawkden. The manor of *Thurstanestuna* is represented by Thurstan's Farm.

HAWSTEAD *(4 m S of Bury St Edmunds)* 'Horz-tid'
Thingoe Hundred. O.Merc. *hald*; AS. *stede* - 'Sloping place' (Skeat) *Haldsteda* (DB). Such a shelving site would describe the position of the parish's eastern part, sloping down to the river Lark. Morley, however, identifies it from its other Domesday spelling of *Hersteda* as 'Place of the army'. Hawstead Place, pulled down in 1827, was from c1460-1610 the seat of the powerful knightly family of Drury (from whom Drury Lane in London is named). Although the house has now gone, the fine moat, described as providing a defence between a castle and a manor house, remains. It may well be that the history of this site goes back to Saxon times and is that referred to in the place-name as being the 'place of the army'. The hamlet of Hardwick means *Herewic*, 'village of the army'.

HELMINGHAM *(9 m N of Ipswich)* 'Hel-me-gum'; 'Hemm-ing-um'; 'Hem-le-gum'
Bosmere & Claydon Hundred. AS. *Helminga ham* - 'The homestead of Helm's people' *Helmingbeham* (DB). The oldest known site in the parish is that of Creke Hall, built in the C12.

HEMINGSTONE *(5½ m N of Ipswich)* 'Hemm-ing-stone'; 'Hamm-ing-stone'; Hemp-st'n'
Bosmere & Claydon Hundred. ?AS. *Hemelinges tun* - 'Hemele's family's farm' *Hamingestuna* (DB). Church Farm shares the natural? hoh on which stands the Saxon church of St. Gregory and may be the original site of the place-name.

HEMLEY *(5½ m S of Woodbridge)* 'Helm-ly' (1764); 'Hemm-lee'
Colneis Hundred. AS. *Helman leah* - 'Helma's clearing or meadow' *Helmelea* (DB). The *l* has been sadly dropped over the last 200 years from the correct and far more pleasing Helmsley and Helmly.

HENGRAVE *(4 m W of Bury St Edmunds)* 'Henn-grave'
Thingoe Hundred. AS. *haema? graef* - The meaning is uncertain. It may be 'Grassy meadow of a man called Hemma' or 'Family burial place'? *Hemegretham* (DB); *Hemegrede* (c1095).

HENHAM *(4 m E of Halesworth)* 'Henn-um'
Blything Hundred. AS. *hean hammae* - 'High homestead or enclosure' *Henham* (DB). Henham is bounded southerly by the river Blyth, on the E and N by a tributary of the same running down from Westhall, and westerly is a (now) slight rivulet, making it a strong stratigic position in Saxon days. It also had a church in Norman times. The long-lost estate of *Warle* (AS. *waer leah* - 'weir (or fish-trap) meadow') was sited in the area, probably running down to the river.

HENLEY *(4½ m N of Ipswich)* 'Henn-ly'
Bosmere & Claydon Hundred. AS. *hean leage* - 'At the high meadow' *Henleia* (DB). The river Finn rises in the middle of the village at Henley Watering.

HENSTEAD with HULVER *(5½ m SE of Beccles)* 'Henn-sted'
Blything Hundred. AS. *hean stede* - 'At the high place' *Henesteda* (DB). Morley states that Henstead's Hall was far more conspicuous in the days before the Kessingland Dam was built across

the Beachfarm marshes dividing the Hundred of Lothingland from that of Blything. The hamlet of Hulver Street which derives its name from the quantity of hulver (provincial name for holly) trees which formerly grew here provided the ford (some five miles inland) through the (then) vast expanse of the Hundred River. Here at Hulver Bridge six ways still meet.

HEPWORTH *(5 m NE of Ixworth)* 'Hepp-uth'
Blackbourn Hundred. AS. *heope (Heppa) worth* - 'Property abounding in wild roses'? or 'Heppa's enclosure'? *Hepworda* (DB).

HERRINGFLEET *(6 m NW of Lowestoft)* 'Hare-in-feet' with pause after 'Hare' *(old)*, 'Herr-en-fleet' without pause *(later)*.
Mutford & Lothingland Hundred. AS. *Herlinga fleet* - 'The shallows of Herla's people' *Herlingaflet* (DB); *Heringfleet* (1340). Refers to the shallow waters or marshes on the E of the river Waveney. The Old Hall stands close above the marshes. St. Olaves takes its name from the priory sited by the bridge. Olave was a Norse King and warrior who was later canonised.

HERRINGSWELL *(3½ m S of Mildenhall)* 'Horn-s'll', *old - earlier generations pronouncing the name straight from its Saxon origins Hyrn or horn*
Lackford Hundred. AS. *Hyrninga wella* - 'Well or Lake of Hyrn's family' *Hyrningwella* (DB). Morley considers this very descriptive of the marshes to the west of Fen Farm having constituted a great Broad running down from the river Lark by way of Tuddenham Mill. Physically this is a difficult part of the county to reconstruct in the C6.

HESSETT *(5½ m SE of Bury St Edmunds)* 'Hedgeset' *(1538)*; 'Hess-et'
Thedwestry Hundred. AS. *hege saete* - 'Settlers at the boundary' *Heteseta* (DB); *Heggeset* (1225). 'At the boundary' may refer to the edge of Bradfield Forest.

HEVENINGHAM *(5½ m SW of Halesworth)* 'Hevv-e-gum'; 'Henn-e-gum'; 'Hinnigum'
Blything Hundred. AS. *Hefaninga ham* - 'The homestead belonging to Hefa's people' *Heueniggeham* (DB). The manor of *Thorpe* in Blything Hundred is thought by the *Victoria County History* to lie in this village. By 1220 it was little more than a berewick of Sibton Manor.

HIGHAM *(5 m S of Hadleigh)* 'Hi-um'
Samford Hundred. O.Merc. *heh*; AS. *hamm* - 'High (or chief) homestead' *Hecham* (1050); *Heihham* (DB). Higham Hall overlooks the junction of the river Brett with the Stour.

HIGHAM *(7 m W of Bury St Edmunds)* 'Hi-um'
Lackford Hundred. O.Merc. *heh*; AS. *hamm* - 'High (or chief) homestead' *Hecham* (1050); *Heihham* (DB). In the S of the parish Higham Hall stands high above a steep rise from the stream running NNW from Denham. This former Gazeley hamlet of Higham Green became a separate parish in its own right in 1894.

HINDERCLAY *(8 m NE of Ixworth)* 'Hin-der Clay' *with emphasis on 'Clay'*.
Blackbourn Hundred. O.Norse. *Hildar clea* - 'Hildr's clay' *Hilderclea* (DB). Hilderclay until the 1381 Poll erroneously replaced the *l* with an *n*. The Domesday *Torp* located here is represented by Thorpe Street.

HINTLESHAM *(4½ m NE of Hadleigh)* 'Hinkle-shum; 'Hint-el-shum'
Samford Hundred. AS. *Hynteles ham* - 'Hyntel's homestead' *Hintlesham* (DB). A fragment of the moat of Timperley's Old Hall which lies just E of Wolves Wood may occupy the site of Hyntel's original homestead.

HITCHAM *(7 m NNW of Hadleigh)* 'Hich-um'
Cosford Hundred. AS. *Hecan ham* - 'Heca's home' *Hecham* (DB).

HOLBROOK *(6 m S of Ipswich)* 'Holl-bruk'
Samford Hundred. AS. *Holum broc* - 'Brook in a hollow or ravine' *Holebroc* (DB). The parish gets its name from a brook which falls into the river Stour at Holbrook Bay near the hamlet of Lower Holbrook, a mile S of the village.

HOLLESLEY *(6 m SW of Orford)* 'Hose-ly'
Wilford Hundred. AS. *Holes leah* - 'Hol's meadow' or 'wood clearing in a hollow' *Holeslea* (DB). The 1086 township of *Culeslea* (Culsey) was a berewick of Hollesley. The place, last recorded as *Culleste* in the 1228 Feet of Fines, may mean 'Cul's meadow' or 'dove's meadow'.

HOLTON *(1 m E of Halesworth)* 'Hole-t'n'
Blything Hundred. AS. *holan tun* - 'Farm in the hollow' *Holetuna* (DB). The farm may have stood near the church which occupies a hollow in the S of the parish.

HOLTON ST MARY *(4½ m SSE of Hadleigh)* 'Holl-t'n'
Samford Hundred. AS. *Holan tun* - 'Hola's farmstead' plus church dedication. *Holetuna* (DB).

HOMERSFIELD (or South Elmham St Mary) *(5 m WSW of Bungay)*
Wangford Hundred. AS. *Hunmaer (Hamar) feld* 'Open land of a man called (Hunbeorht) (Skeat) or Hamar' (Morley) *Humbresfelda* (DB); *Humeresfeld* (1250). The church of St Mary stands on an ancient platform above the Waveney and is believed by Morley to have been erected upon the site of a temple to the pagan Saxon god Hamar. Domesday shows thirty acres owned by Bury Abbey under *Linburna* in the extreme SW of the parish where the Waveney's tributary was expanded into a pool (gaelic. *linn*). It is now represented by Limbourne Common.

HONINGTON *(3 m NW of Ixworth)* 'Honn-in-ton' or 'Hynnington'
Blackbourn Hundred. Dan. *honning*; AS. *tun* - 'Honey farm' *Hunegetuna* (DB).

HOO *(4 m SSW of Framlingham)* 'Hoo-o', *distinctly in two syllables*
Loes Hundred. AS. *hoh* - 'Spur of a hill' *Ho* (c1050); *Hou* (DB). Obviously an important place in Saxon times when the entire village was embraced by one large manor of three caracutes (360 acres) under Ely Abbey. Hoo Hall stands on a bold hill-spur above the Deben valley and is the likely earliest settlement site.

HOPTON BY THE SEA *(5 m N of Lowestoft)* 'Hopp-t'n'
Mutford & Lothingland Hundred. AS. *hop tun* - 'Place at the crescent bay' *Hopetuna* (DB). While the meaning is given by Skeat as 'Farm in a valley (or coast) recess', Morley believes the name to come from the Anglo-Saxon *hop*, a circle, (medieval *hoope*) meaning, in this case, a crescent bay. A similarity can be drawn with Weybourne on the Norfolk coast. The saying goes "He, who would Old England win, must at Weybourne Hoope begin." This bay has now been washed near flat. If a similar bay once existed here it has also been eroded away as the coastline is now virtually straight. It is further S in Blything Hundred where Domesday has two entries (at present unlocated) for *Hoppetuna* and *Opitun*, possibly sited about where Low Farm now stands.

HOPTON *(7 m NNE of Ixworth)* 'Hopp-t'n'
Blackbourn Hundred. AS. *hop tun* - 'Farm in a valley recess' *Hopetuna* (DB). Although strategically a strong position (forming almost an island of boulder clay surrounded by marshes

and approached by the narrow neck of Weston Fen) it appears to have been late settled and never a fortified place with no existing earthwork and a church with no evidence of early Saxon architecture.

HORHAM *(5 m SE of Eye)* 'Hor-rum'
Hoxne Hundred. AS. *horu hamm* - 'Dirty or muddy homestead' *Horham* (c950); *Horham* (DB). The name of this parish has been uncorrupted from the first. The entry of two of its parcels in Domesday under Hartismere Hundred looks as though it extended further W than at present. Horham lies on boulder-clay hence the muddy district.

HORRINGER (or Horningsheath) *(2 m WSW of Bury St Edmunds)* 'Horren-jer'.
'Horningsheath' *is still used occasionally*
Thingoe Hundred. AS. *horninges heorth* - 'Horn's son's house' *Horningeserda* (DB). There were originally two parishes here, *Horningsheath Magna* and *Horningsheath Parva*. The name was already being slurred to the clipt form *Horenser* by the 1204 Feet of Fines. The hamlet of *Horsecroft*, meaning 'paddock', is not mentioned before 1381.

HOXNE *(3½ m NE of Eye)* 'Hox'n'
Hoxne Hundred. AS. *Hoxena* - 'Settlement of the Hoxan (tribe)'? *Hoxana* (DB). Heckfield (AS. *haccan* - 'to gash' and Dan. *hekelen* - 'to tear assunder') Green stands on the southern hill once covered by the more extensive Hoxne Wood, the alleged site where King Edmund was slain by the Danes in 870 A.D.

HUNDON *(6 m ENE of Haverhill)* 'Hunn-d'n'
Risbridge Hundred. AS. *Hunan denu* - 'Huna's valley' *Hunedana* (DB). The parish lies in a dale of gently undulating country that extends right across from Poslingford to Bradley.

HUNSTON *(3 m SE of Ixworth)* 'Hunts-t'n'
Blackbourn Hundred. AS. *Hunteres tun* - 'Hunter's farmstead' *Hunterstuna* (DB). Hunter, a man so nicknamed from his prowess in the chase, likely settled on the site of the old hall which was doubly moated and destroyed by fire in 1917.

HUNTINGFIELD *(4 m SW of Halesworth)* 'Hunt-'n-field'
Blything Hundred. AS. *Huntinga feld* - 'Clearing belonging to Hunta's family' *Huntingafelda* (DB). St. Marys Church lying in the valley of a tributary of the river Blyth retains Saxon work.

ICKLINGHAM *(3½ m ESE of Mildenhall)* 'Ick-e-gum'
Lackford Hundred. O.Merc. *Icelinga ham* - 'Icel's family's home' *Ecclingaham* (DB). Icklingham comprises two united parishes, All Saints and St James, each with its own church.

ICKWORTH *(4 m SW of Bury St Edmunds)* 'Icker *(1900)*
Thingoe Hundred. AS. *Iccan worth* - 'Icca's property' *Ikewrth* (c950).

IKEN *(3½ m W of Aldeburgh)* 'I-ken' *as in 'liken'*.
Plomesgate Hundred. Latin. *Iceni* - 'The place of the Iceni' *Ykene* (1212). Though it is identified by Skeat as 'Ica's stream', Morley with his local knowledge supported the views of Isaac Taylor (1878) and Maclure (1910) in thinking it 'The place of the Iceni'. In 654 AD, the year of Penda's victory at the Battle of Blythburgh, "Botolf began to timber the minster at Ycean-ho". The place was likely to have been Yarn Hill, and he identifies the *ho* of *Ycean ho* as 'the bluff jutting N into the river Alde; at the root of which bluff the church stands.' Whilst not apparently named in Domesday, Morley suggests it may be the *Udeham* (Woodham) at present unidentified and placed between Aldeburgh and Campsey.

ILKETSHALL - (ST ANDREW, ST JOHN, ST MARGARET, ST LAWRENCE)
(4 m SE of Bungay) 'Ill-kett-shawl'
Wangford Hundred. Norse. *Ulketeles*; AS. *healh* - 'Ulfcytel's shelter' plus church dedications. *Ilcheteleshala* (DB). The Mount, situated in a copse at Manor Farm, ½ mile ENE of the church, is a late C11 or early C12 motte and bailey castle occupied by the Ilketshall family between the C12-C15, but it may have earlier origins and be the original Ulfcytel's 'shelter'. However, as with the South Elmhams, Ilketshall comprises a number of parishes divided by their church dedication so a more obscure ecclesiastical meaning may apply. Here in Ilketshall (tentatively) may be placed the Domesday *Catesfella* (AS. *Ceattan feld* - 'Ceatta's clearing').

INGHAM *(4 m N of Bury St Edmunds)*
Blackbourn Hundred. O.Norse. *eng*; AS. *ham* - 'Meadow home' or 'Inga's homestead' *Ingham* (DB). It has recently been suggested that the first element of the name may refer to a member of the ancient Germanic tribe called the *Inguiones*.

IPSWICH 'Ipps-itch' *(North)*, 'Ippsedge' *(West)* and even "Hippsetch'
AS. *Gipes wic* - 'Gipi's (Gippy's) enclosure (harbour)'? *Gipeswic* (c975). Although the town actually stands on the river Gipping, which only takes the name Orwell from the spot where it becomes tidal (Stoke Bridge), all experts agree that the first element of the town's name *Gip* is, in fact, taken from a personal name, i.e. 'a man called Gipi'. The initial *G* was sounded as a *y* and consequently drops out entirely leaving *Ypeswic*. *Gipi's* son would, however, be called *Gipping*. With its pottery industry producing 'Ipswich ware' from at least c625 AD, the town may have been one of the largest known post-Roman trading centres in NW Europe.

IXWORTH *(6½ m NE of Bury St Edmunds)* 'Ix-uth'
Blackbourn Hundred. AS. *Gixan wearth* - 'Gixa's property' *Gyxeweorde* (c1025); *Giswortha* (DB). There is evidence of early Saxon occupation of the Roman villa site SE of the village.

IXWORTH THORPE *(1½ m NW of Ixworth)* 'Ix-wuth Thorpe'
Blackbourne Hundred. O.Fris. *thorp* - '(Ixworth's) secondary settlement' *Torp* (DB); *Ixeworth Thorp* (1305); *Ixeworthethorp* (1327). The village was formerly a manor of Ixworth Priory which, at the Dissolution, passed with Ixworth to Richard and Elizabeth Codington. Placed in Domesday in both Bradmere and Blackbourn Hundreds, it was never associated by name with Ixworth until *Thorp de Ixworth* (1280).

KEDINGTON *(3 m ENE of Haverhill)* 'Ketton' *(1764/1885)*; 'Kitt'n' *(1919)*; 'Kedd-'n-t'n'
Risbridge Hundred. AS. *Cydan Tun* - 'Farmstead of a man called Cydda' *Kydington* (1043-5); *Kidituna* (DB).

KELSALE *(1½ m N of Saxmundham)* 'Kels-ull'
Hoxne Hundred. AS. *Ceoles healh* - 'Ceol's shelter' *Keleshala* (DB). Kelsale Hall lies in a sheltered position in the valley of a tributary of the river Fromus. It incorporates the smaller parish of Carlton and is now known as Kelsale-cum-Carlton.

KENTFORD *(4½ m E of Newmarket)* 'Kenn-ford'
Risbridge Hundred. AS. *Cyneta forda* - 'Ford over the river Kennet'. Not in Domesday under Suffolk and so close to the present border with Cambridgeshire that it may be included there. *Cheneteforde* (C11); *Kenteford* (1275). The prehistoric Icknield Way (B1506) fords the river Kennet just W of the street and church, giving the parish its Saxon place-name.

KENTON *(2½ m NNE of Debenham)* 'Kent-'n'
Loes Hundred. Celt. *cefn*; AS. *tun* - 'Farm on the hill-ridge' *Chenetuna* (DB). Skeat considers its meaning 'Cena's farm', but Morley is confident that the place-name is evident in the long ridge of the western hill running parallel with the stream from Kenton Lodge to the church.

KERSEY *(2 m NW of Hadleigh)* 'Carsey' *or* 'Kay-sy'
Cosford Hundred. Lat. *chorus* or AS. *chores*; AS. *ig*. 'Assembly Island' *Caeresige* (c995); *Careseia* (DB). Although many authorities give the meaning 'Island where cress grows', Morley feels it relates directly with Cosford Hundred - 'Island of the Cors' and 'Ford of the Cors'. The village stands in an elevated position running down to the valley of the river Brett. The river has now taken its name from Brettenham but if, as seems likely, it was originally called Cors, it gave rise to Cosford Hundred and the lost Domesday township of *Corsfield* which lay in the neighbouring Babergh Hundred. Cosford Hall in Hadleigh may well be the site of the later Hundred Court since it immediately flanks the stream, but an earlier and original Cosford Hall is shown on early OS maps as that island in the Brett which gave name to both Kersey and the Hundred. Here too Kirby names Cosford Bridge.

KESGRAVE *(3 m NE of Ipswich)* 'Kez-grev'
Carlford Hundred. Norse. *Ketilles*; AS. *graef* - 'Ketill's grave' *Gressegraua* (DB); *Kersigrave* (1231). Morley considers it a Norse personal name connected with one of the burial mounds in the parish and taking it back to the C9.

KESSINGLAND *(5 m SSW of Lowestoft)* 'Kess'n-lund''
Mutford & Lothingland Hundred. AS. *Casinga land* - 'Casa's family's land' *Kessingalanda* (DB).

KETTLEBASTON *(7 m NNW of Hadleigh)* 'Kettle-bar-st'n'
Cosford Hundred. Norse. *Ketelbarnes*; AS. *tun* - 'Ketel's child's farmstead' *Kitelbeornastuna* (DB). Always shown as Kettlebarston until William White omitted the *r* in his 1844 directory. Domesday shows the entire township represented by a part of the manor of *Manetuna* which was then six furlongs in length and four in breadth. The modernised C17 Kettlebaston Hall stands by the church on the site of previous halls.

KETTLEBURGH *(2½ m SSW of Framlingham)* 'Kerry-brah', *now* 'Kettle-borough'
Loes Hundred. Norse. *Ketilan byrig* - 'Ketila's hill or tumulus' *Ketelbiria* (DB). Broomfield, *Brumfella* (DB), was an (at present unlocated) estate here.

KIRKLEY *(1 m SSW and now forming a part of South Lowestoft)* 'Kirk-ly'
Mutford & Lothingland Hundred. O.Norse. *kirkja*; AS. *leah* - 'Clearing near or belonging to a church' *Kirkelea* (DB). The modern church was in ruins in 1735 (Kirby) and stands on the NW extremity of a boulder-clay peninsula running N; its position relates it, not with the coast but with the Carlton Colville stream at Kirkley Run, a ford-name now extended to a group of adjacent cottages, called in 1274 *Wicfleet* i.e. 'shallow creek'.

KIRTON *(6 m S of Woodbridge)* 'Kur-t'n'
Colneis Hundred. O.Norse. *kirkja*; AS. *tun* - 'Farmstead with a church' *Kirketuna* (DB); *Kirkton* (1783). Again we have the rare occurrence of a parish which anciently held a Norse church, though the earliest parts of the present building only date from the C14 and C15.

KNETTISHALL *(8 m NNE of Ixworth)* 'Knatteshall' or 'Netts-ull'
Blackbourn Hundred. ?Norse. *Knattes*; AS. *healh* - 'Knotte's shelter' *Gnedeshalla* (DB). The

place-name may apply instead to a very pestilent biting gnat still prevalent over all the Brecklands of Norfolk and Suffolk.

KNODISHALL-CUM-BUXLOE *(4 m SE of Saxmundham)* 'Noddy-shul'
Blything Hundred. ?AS. *Cnotes healh* - 'Cnot's shelter or corner of land' *Cnotesheala* (DB). Buxloe (AS. *bucces hlaw* - 'deer's burial mound') was consolidated with Knodishall in 1721. No reference to Buxloe appears before *Buckyslowe* (1247). The Poll Tax of 1381 shows a full township at *'Buxlowe in Blything'* with 30 inhabitants. A part of the church's circular Norman tower alone remains at Knodishall Green.

LACKFORD *(8 m NW of Bury St Edmunds)*
Thingoe Hundred. AS. *lacu forda* - 'Ford over a stream' *Lecforda* (DB). The pre-historic Icknield Way crosses the river Lark by the ford which gave name to the parish. This great Domesday manor of 600 acres with two water mills then gave name to Lackford Hundred. Whilst it is true that the parish is actually listed in Thingoe Hundred, the river crossing lies exactly astride the hundred boundaries of Lackford, Thingoe and Blackbourn.

LAKENHEATH *(5 m N of Mildenhall)* 'Lake-'n-heath'
Lackford Hundred. AS. *lacinga hythe* - 'Hithe of the stream dwellers' *Lacingahithe* (945); *Lakingahethe* (DB). 'Heath' comes from the Anglo-Saxon *hyth* meaning 'a landing place for boats'. The active waters of the Old Fen Sea later partially dried out and (with a programme of drainage) became the fens of today. A highly interesting but controversial suggestion as to the origins of the parish has been put forward to satisfy the puzzle over the fact that the northern parish and county boundary (with Norfolk) - fails to follow the line of the Little Ouse River for its whole length. The boundary is said to have been set out by the order of Abbot Brithnoth *c*974 AD at a time when a large area of Lakenheath parish was still covered by a now extinct lake. An area of land just to the NW is called Redmere Fen, probably derived from the Old English word *Hreod* meaning reed, hence Reed Mere. Examination of the surface of the land showed shell-marl covering an area of 2,200 acres which would have made it the largest lake in England after Windermere. It probably came into existence 500-100BC and, being shallow, its margins were subject to drying out in Summer.

LANGHAM *(3 m E of Ixworth)* 'Lang-um'
Blackbourn Hundred. AS. *lang hamm* - 'Long homestead or enclosure' *Langham* (DB). Township names were not parochial, hence the nearly square figure of the present parish has no bearing upon its place-name which probably refers to the original manor embracing the hall and the area between Castle Ditches and the river Brett's tributary stream.

LAVENHAM *(6 m NNE of Sudbury)* 'Lavv-num'
Babergh Hundred. AS. *Lafan ham* - 'Lafa's homestead' *Lauenham* (DB). The site of the great hall of John de Vere, 13th Earl of Oxford (perhaps Lafa's original Saxon site) is marked by a large wooded moat standing in an elevated position NW of the present town. By 1525 Lavenham, an obscure village in the early Middle Ages, had raised itself - as one of the main East Anglian centres for the thriving cloth industry - to rank as the 20th most wealthy town in England.

LAWSHALL *(6 m S of Bury St Edmunds)* 'Lorz-'l'
Babergh Hundred. AS. *lagu sele* - 'Dwelling by the lake' *Lawessela* (DB). Suffix is not *hale* (hall) but *sele* meaning dwelling place. Morley suggests that this original place was beside the

stream near Coldham Hall where geological surveys show that tracts of alluvium lie E and W of The Warbanks which may well have been lakes 2,000 years ago. With the dense timber on either flank on the flat heavy boulder clay, and the ramparts of Warbanks, this would make a notable military post.

LAXFIELD *(6 m NE of Framlingham)* 'Laax-vul' *or* 'Lax-ful'
Hoxne Hundred. Norse. *Laxan*; AS. *feld* - 'Leaxa's clearing' *Laxefelda* (DB). This place was settled by a Viking and apparantly named after him by the Saxons. The Domesday manor of *Stadhaugh* (AS. *stede healh* - 'farmstead in a sheltered position') is represented by Stadhaugh Manor Farm N of the village.

LAYHAM *(1½ m S of Hadleigh)* 'Li-um'
Cosford Hundred. ?AS. *hleow hamm* - 'Sheltered enclosure' *Hligham* (c995); *Leiham* (DB). The village lies in the valley of the river Brett.

LEAVENHEATH *(1 mile NW of Nayland)* 'Levv'n-heath'
Leavenheath, AS. *Leofwines haeth* - 'Leofwine's heath', is a parish created on 13 January 1863 from parts of Stoke by Nayland, Wissington, Assington, Nayland and Polstead. Leaden (Leofwine's) Heath was a broad open space on Faden's 1783 map between Wissington Grange and Leaden Hall, now enclosed.

LEISTON 'Lay-st'n'
Blything Hundred. AS. *Leofstanes tun* - 'Leofstan's farm' *Lehtuna* (DB). The three Domesday entries in Colneis Hundred under *Leofstan(t)estuna* apply to the now lost township of *Leyston* in Trimley St. Mary.

LETHERINGHAM *(3½ m SW of Framlingham)* 'Leth-er-gum'
Loes Hundred. AS. *Leoderinga ham* - 'Leodhere's family's home' *Letheringaham* (DB). Letheringham Hall occupies an ancient moated site which may once have provided Leodhere's family's home.

LEVINGTON *(7 m SE of Ipswich)* 'Lennington' or 'Lever-ton'
Colneis Hundred. AS. *Leofan tun* - 'Leofa's farmstead' *Leuentona* (DB). Levington includes the Domesday vills of *Stratton* - 'farm on or alongside the Roman road', now represented by Stratton Hall; *Strudestun* - 'farm of the robber'; *Culuerdestuna* - '(Lady) Culfre's farm', and *Guthestuna* - 'Guth's farm'. The Domesday *Isteuertona* - 'Isetheof's farm' is thought by Copinger to also lie in this parish. Morley points to the comparative density of population in Colneis Hundred at this time when the dearth of timber upon its light soil and the presence of the Orwell and Deben estuaries on either side resulted in the development of a number of small but active hamlets.

LIDGATE *(7 m SE of Newmarket)* 'Lid-gett'
Risbridge Hundred. AS. *hild geat* - 'Place at the swing or clapper-gate' (AS. *hlid* means hinged lid of a box) *Litgata* (DB). Morley hypothesises that if we take the castle here to have originally been a fort (as the local poet John de Lydgate suggests in his poem: 'Lydgate . . By olde time a famous castel towne, In Danes time it was beate down . .') perhaps built (by King Anna?) to repel a hostile force's outflanking movement along the southern side of the Devil's Dyke into Suffolk; to any such tactic this fort would present a 'side bar'. Although not marked as such on any map, it has been suggested that the brook following the river Kennett was once called the *Lydd*, a common Celtic river-name found in at least four English counties.

LINDSEY *(4½ m NW of Hadleigh)* 'Linns-y'
Cosford Hundred. AS. *Lilles eg* - 'Island in the marshes of a man called Lill' *Lealeseia* (c1095); *Lelleseye* (1263). There is little doubt about the placing of Lill's 'island in the marshes'. Lindsey Castle now reveals nothing but an irregular series of mounds and dykes, obscured by trees and undergrowth, lying in the bottom of a narrow boulder-clay valley which is still very marshy in its eastward course, to the ravine, cut through glacial gravel to the London clay in which lies Kersey and beyond it to the river Brett. The site was later occupied by a motte and bailey castle mentioned in the C12 and probably built by the powerful de Cokefeld family.

LINSTEAD MAGNA *(4½ m WSW of Halesworth)* 'Linn-sted Magna' *(never 'Great')*.
Blything Hundred. AS. *lin stede* - 'Place where flax is grown' *Linestede* (DB). Our two townships of Linstead Magna and Linstead Parva were one through medieval times. A law was passed in the reign of Henry VIII that every parish must grow flax, quite irrespective of soil. The plant was not indigenous to England, but imported from the East, revealing the Saxon's overseas commerce.

LINSTEAD PARVA *(4 m WSW of Halesworth)* 'Linn-sted Parva' *(never 'Little')*.
Blything Hundred. AS. *lin stede* - 'Place where flax is grown' *Linestede* (DB). Linstead Parva was also known as *Lower Linstead* and *Linstead Chapel*.

LITTLE BEALINGS *(3 m SW of Woodbridge)* 'Liddle Ballens or Beelens'
Carlford Hundred. AS. *Beolingas (parva)* - 'Settlement of the Beola (tribe)' or 'Settlement by the funeral pyre' *Belinges* (DB). *(see Great Bealings)*

LITTLE BLAKENHAM or Blakenham Parva *(4 m NW of Ipswich)* 'Liddle Blake-num'
Bosmere & Claydon Hundred. AS. *Blacan ham (parva)* - 'Blaca's homestead or enclosure' *Blac(he)ham* (DB). Blakenham Magna and *Blakeneham Parva* were one parish at Domesday with one church. Blakenham Parva church was later enumerated under Nettlestead.

LITTLE BRADLEY *(6 m N of Haverhill)* 'Liddle Bradd-ly'
Risbridge Hundred. AS. *braden leah (parva)* - 'Broad meadow' *Bradeleia* (DB).

LITTLE FINBOROUGH *(2½ m SW of Stowmarket)* 'Finn-brah' or 'Fimm-brah'
(See Great Finborough). At Domesday Finborough was but a single entry. The part of it in Bricett manor which descended, not with Finborough, but with Buxhall represents the present village of Little Finborough.

LITTLE GLEMHAM or Glemham Parva *(7½ m NE of Woodbridge)* 'Liddle Glemm-un'
Plomesgate Hundred. ?AS. *Glaem hamm* - The meaning of this place-name is obscure and several authorities have had their say without real conviction. Skeat favours 'Gleam-enclosure' (from the AS. *glaem* - 'brightness' i.e. 'a sunny situation' comparing our other village Glemsford - 'Ford through the Gleam River'), whilst Reyce's Breviary of 1618 claims that the river Alde was earlier termed the Gleme 'which cometh from Rendlesham (Rendham?) and both the Glemhams' *Glaimham* (DB). Morley places the Domesday *Beuresham* here. The bridge between this parish and Blaxhall is called Beversham Bridge and among Chantry Lands were "the manor of Beversham and lands called Beversham in Glemham Parva."

LITTLE LIVERMERE or Livermere Parva *(4 m NW of Ixworth)* 'Livv-mere'
Blackbourn Hundred. AS. *laefer mere* - 'Mere of yellow-iris' *Litla Livermera* (DB); *Lyuermere Parua* (1327). The lake becomes obvious when we notice that there a tributary of the river Lark

rises in a broad splay, forming an ornamental mere in Livermere Park. The prefix is a misnomer as both parishes are large; Great Livermere occupies 1500 acres and Little Livermere, 1400. They are also in different Hundreds with the line of demarkation running down the centre of the lake emphasising its early importance.

LITTLE SAXHAM *(4 m W of Bury St. Edmunds)* 'Liddle Saxum'
Thingoe Hundred. O. Merc. *Saxan ham* - 'Seaxa's hopmestead' *Saxham Parva* (DB).

LITTLE STONHAM or Stonham Parva or Stonham Jerningham *(6 m ENE of Stowmarket)* 'Little Stonn-um'
(See Earl Stonham). The Jerningham family were anciently manorial lords here for many years and during this period the parish was named after them.

LITTLE WALDINGFIELD or Waldingfield Parva *(4½ m NE of Sudbury)*
'Wonnerfeld' *or* 'Wannerful'
Babergh Hundred. O.Merc. *Waldingafeld* - 'Walda's family's clearing' *Wealdingafeld* (c995); *Waldingefelda* (DB).

LITTLE WENHAM or Wenham Parva *(6½ m SW of Ipswich)*
Samford Hundred. AS. *Wenan ham* - 'Wena's Homestead' *Wenham* (DB). Probably a single township until after the Conquest. It was first divided into Magna and Parva in the 1327 subsidy.

LITTLE WHELNETHAM or Welnetham Parva *(4 m SE of Bury St Edmunds)*
'Weltham' *or* 'Little Well Neeth-um'
Thedwestry Hundred. ?AS. *hweol Witan ham* - 'Wita's circular home?' or 'Enclosure frequented by swans near a water-wheel (or some other circular feature)?' *Hvelfiham* (DB); *Weluetham* (1170). One township at Domesday, the place-name may refer to the ancient mere at Sicklesmere, a hamlet to the N of Great Whelnetham parish (and/or a possible ancient watermill).

LITTLE WRATTING or Wratting Parva *(2 m NE of Haverhill)* 'Tallow Wratting' *or* 'Ratt'n'
Risbridge Hundred. AS. *Wraettinga* - 'Of Wraett's family' *Wratinga* (DB).

LONG MELFORD *(3 m NW of Sudbury)* 'Mell-fud', *the 'Long' is not locally used*
Babergh Hundred. ?AS. *maelan forda* - 'Maela's ford' or 'Ford by a mill' *Melaforda* (DB). 'Long' did not prefix the name until some time after 1327. There were two ancient fords across the Chad Brook on the course of the Roman road. One by the site of the mill at the northern end of the street (the last on the site demolished in the 1950s.) relates to the place-name, and the other by Ford Hall in the hamlet of Bridge Street on the northern parish boundary. Kentwell *(Kanewella)* Hall is from the AS. *Cenan wella* - 'Cenan's well'. The well lies just S of the hall.

LOUND *(5½ m NNW of Lowestoft)* 'Leund' *as the Christian name 'Hugh'*
Mutford & Lothingland Hundred. O.Norse. *lundr* - '(sacred?) grove' *Lunda* (DB). Morley suggests it as probably sacred; a grove where all the Norsemen of the neighbourhood assembled to pay homage to their gods in the days before Christianity. The fragment of Great Wood on the southern boundary is all that remains of ancient woodland and may form the 'grove' from where the parish obtained its Domesday name.

LOWESTOFT 'Low-es-toff'
Mutford & Lothingland Hundred. AS. *Hlothuwiges toft* - 'Ludwig's knoll' *Lothu Wistoft* (DB);

Lothewistoft (1212). Morley suggested the knoll may have applied to the rounded prominence of the Old Town cliff near the lighthouse. The original settlement of Lowestoft comprised one narrow street (High Street) about 1½ miles in length with a few smaller streets and lanes (known locally as scores) branching off, and was confined to the northern side of Lake Lothing which divided it from the extensive Kirkley Heath to the S.

MARKET WESTON *(7 m NNE of Ixworth)* 'Mar-kut Wes-t'n'
Blackbourn Hundred. AS. *west tun* - 'West farmstead or village' *Westuna* (DB). Robert Hovel obtained a grant for a market in 1264 but the township did not aquire the name 'Market' until the C14. It is difficult to suppose which important centre lay east of this Weston but it may have been Botesdale.

MARLESFORD *(5 m SSE of Framlingham)* 'Molls-feh' *(early)*; 'Marlsfud' *(later)*
Loes Hundred. O.Norse. *Maerles*; AS. *forda* - 'Maerl's ford' *Merlesforda* (DB). The parish obtains its name from an ancient ford. This ford through the river Ore lay on the great highway (A12) between Beccles and Ipswich (Yarmouth, likely, did not then exist) in the S of the parish and approached from the N by Hollow Lane.

MARTLESHAM *(2 m SSW of Woodbridge)* 'Martel-shum'
Carlford Hundred. O.Norse. *morthr*; AS. *ham* - 'Homestead by a woodland clearing frequented by martens' *Merlesham* (DB); *Martlesham* (1254). It appears likely that the first settlement in the area was on the S bank of Martlesham Creek where the isolated church and hall now stand. It is known that the hall site is an early one and the place-name meaning would certainly apply at this spot - a clearing with a maze of footpaths by the creek between the ancient woodland of Sluice Wood and Lumber Wood. Across the creek was the ancient royal vill of Kingston. *Barchestuna* was then a small manor of one carucate (120 acres) which emerges under the name Barkestone in 1346 but did not persist as it was unknown as manorial to Copinger. *Barking* is shown to mean 'at the birch-grove' so here is the same AS *beorc* with the distinction of 'birch farm'. There are various references to Barkestone of which the following should suffice to place it: 1442: 'One acre . . . in Barkestone hamlet of Martlesham formerly Robert Genew.' 1460: 'Thomas Buttler . . in Barkeston in Martlesham. 1552: 'My tenament called Butteleers . . . in the hamlet of Barkeston in Martlesham' The 120 acres of *Prestetuna* - 'Priest's Farm' in Carlford Hundred also lay within this parish.

MELLIS *(3½ m W of Eye)* 'Mell-us'
Hartismere Hundred. AS. *(?Drag)meles or melu-hus ?* - An uncertain origin. It may mean '(?Drag)mel's farm' or perhaps 'meal-house (water mill)?' *Melles* (DB).

MELTON *(1 m NE of Woodbridge)* 'Mell-t'n'
Wilford Hundred. Norweg. *mel*; AS. *tun* - 'Farm by the sandbank' *Meltune* (c1050); *Meltuna* (DB). Though there is considerable debate over the meaning of the place-name, Morley suggests it thus and points to the sandbank along the water-course 'as apparent today (1950) as it was when the Vikings sailed up the Deben to harry the land a thousand years ago'.

MENDHAM *(8 m SW of Bungay)* 'Mend-um'
Hoxne Hundred. AS. *Myndan ham* - 'Mynda's homestead' *Myndham* (c950); *Mendham* (DB).

MENDLESHAM *(7 m NNE of Stowmarket)* 'Mend-le-shum' *or* 'Menn-el-shum'
Hartismere Hundred. AS. *Myndeles ham* - 'Myndel's homestead or village' *Mundlesham* (DB).

Mendlesham's pre-Conquest past is most elusive and intriguing with mere suggestions of a royal residence, ring, crowns, even 'Dagobert, one of the East Anglian kings' is said to have lived here. There is a suggestion that the church's first dedication was to King Aethelbeorht and in the present St. Mary's is certainly interred an unnamed Bishop of Norwich.

METFIELD *(8 m SSW of Bungay)* 'Mett-field'
Hoxne Hundred. AS. *mead feld* - 'Place cleared in the forest to make meadowland' *Medefeld* (1214). Until the 1327 Subsidy, *Medefeld* had been regarded as a mere hamlet of Mendham and centuries later in 1764 Kirby stated that "Metfield is sometimes called a Chapel of Mendham of which parish it was a part in the time of Edward the Confessor and Bishop Aylmer".

METTINGHAM *(2 m E of Bungay)* 'Mett-en-gum'
Wangford Hundred. AS. *maetinga ham* - 'The homestead belonging to Maete's family' *Metingaham* (DB). This was the home of the sons of Maete whose personal name comes from the adjective *maete* meaning 'moderate, small, poor', suggesting he was a man of mediocre capabilities or stature. Wainford (AS. *waegn ford* - 'waggon ford') on the boundary with Bungay, which once provided an important crossing of the Waveney and gave its name to the Hundred, is now merely the site of Wainford Mills, but it was still a township as late as the reign of Edward I when we find a grant of land in "Wangford by Mettingham".

MICKFIELD *(3 m WSW of Debenham)* 'Mick-ful'
Bosmere & Claydon Hundred. AS. *Micelan feld* - 'At the broad clearing' *Mucelfelda* (DB). Morley notes that this settlement, between two Roman roads forking northwards from Sharpstone in Barham, was at a site already cleared, suggesting this was not the first settlement which is likely to have been occupied before Saxon times. It was more probably at the church than at its more northern Hall.

MIDDLETON-CUM-FORDLEY *(4 m NNE of Saxmundham)* 'Middle-t'n'; 'Ford-ly'
Blything Hundred. AS. *middel tun* - 'Middle farmstead' *Mideltuna* (DB). Middleton was situated along an ancient packway and its place-name doubtless refers to it being the middle township along this route between Dunwich and Kelsale. This is shown to have run direct from near the erstwhile Hundred Mere in Kelsale to Eastlands here; then turned at right angles northwards, crossed the Minsmere River at Wrackford Run (ford), and thence passed straight across Westleton Heath to Dunwich, making Middleton 3½ miles from both ends. The packway was in general use from the days of Dunwich's opulence till at least 1481. Fordley, AS. *ford leah* - 'meadow beside the ford', is a lost township of which the church's debris lies just S of Middleton's church in the same graveyard. The ford in question is Wrackford Run which, in 1900, was 'unbridged and a most delightfully wild, lonely and romantic spot, gay with burgeoning wild flowers and sunshine'.

MILDEN *(4 m SSE of Lavenham)* 'Mill-d'n'
Babergh Hundred. AS. *Meldinga* - 'The place of Melda's family' *Mellinga* (DB); *Meldinges* (c1130). The village was anciently known as *Melding*, and in the C18, *Milding*.

MILDENHALL 'Mill-d'n-all'
Lackford Hundred. AS. *Mildan hale* - 'At Mild(burh)a's shelter' *Mildenhale* (c1050); *Mitdedehalla* (DB). Morley questions if Mildenhall could have derived its title from that Saint Mildburh who died in 675 AD? She was niece of King Wulfhere of Mercia and grand-niece of the East Anglian King Anna. And did this Mercian princess found and inhabit as its first Mother may

Superior some unrecorded monastery at 'Mild(burh)a's shelter' beside the Lark at High Town?

MONEWDEN *(5½ m SW of Framlingham)* 'Moneden', 'Mun-ey-don'
Loes Hundred. Celtic. *Mynydd den* - 'Deep wooded valley with bare crest' *Munegadena* (DB). Such a place would be the deep valley here through which runs tributaries of the river Deben.

MONKS ELEIGH *(6 m NW of Hadleigh)* 'Mungs-e-ly'
Babergh Hundred. AS. *Illan leah* - 'Illa's clearing in the wood' *Munegadena* (DB); *Monekesillegh* (1304). This was one single entity till the year 991, and Illa was most likely a monk as the manor was given to the monks of Canterbury by Brithnoth, Earl of Essex, who lost his life in 991 AD fighting the Danes in the Battle of Maldon. It was then called Monks Eleigh. After the Dissolution it was divided and given to the Dean and Chapter of Canterbury. The other half became Brent Eleigh.

MONK SOHAM *(3 m NE of Debenham)* 'Mungs-soo-um'
Hoxne Hundred. AS. *sae ham* - 'enclosure by a hollow' *Saham* (DB). Monks refers to Bury Abbey. It was *Sah'm Monach* in 1340 and *Soham monachorum* in 1381. 'The township rises commandingly upon the crest of the Deben valley. Visible for a considerable distance this was just such a site as the dominant Danes of the C11 would select.

MOULTON *(4 m E of Newmarket)* 'Moll-t'n'
Risbridge Hundred. AS. *Mulan tun* - 'Mula's farmstead' *Muletuna* (DB). Stonehall (or White Hall) was the earliest known building here, standing in front of a circular mound which on excavation was found to be surrounded by a stone wall. Moulton Manor, built to replace Stonehall (burned down 1921) was erected *behind* the mound.

MUTFORD *(5 m ESE of Beccles)* 'Mutt-fud'
Mutford Hundred. AS. *mutha forda* - 'Ford at which moots were held' *Mutford* (DB); *Muthford* (1263). "Villages. like kingdoms, have their period of prosperity and decay" states Suckling, "this now obscure parish was of suffcient importance in Saxon days to give name to the Hundred." Four footpaths meet at a footbridge over the Hundred River S of Marsh Lane Farm, the possible site of the ancient ford which formed the meeting place of the Hundred. There are two other possible meanings for the place-name; both very appropriate. The AS *mutha* also means either 'the mouth of a river' - in this case the Hundred River, then a broad body of water across the much later Latimer Dam, or 'a place where two streams joined' - beyond the erstwhile estuary we find that the formerly important road over Hulver Bridge in Mutford crosses a twin ford through the Hundred and a second small river within a few yards of each other.

NACTON *(3 m SE of Ipswich)* 'Nag-t'n' *or* Nakk-t'n'
Colneis Hundred. O.Norse. *Nakka;* AS. *tun* - 'Nakki's farmstead' *Nachetuna* (DB). Morley states that the Danish *knager* means to crack so Nakka may have been a nick-name given to a man perhaps 'given to the habit of knacking skulls!'

NAUGHTON *(5 m N of Hadleigh)* 'Now-t'on'
Cosford Hundred. AS. *atten awal tun* - 'At the orchard farm'? *Nawelton* (c1150). Not a parish at Domesday when it was likely all forested and a part of Whatfield, its flat, waterless countryside unlikely to have attracted early settlers. Along with the apparently late date for its church and spelling it would appear to have Norman rather than Saxon origins. It was united with Nedging in 1934 to form Nedging-with-Naughton.

NAYLAND *(9 m SE of Sudbury)* 'Nay-lund'
Babergh Hundred. AS. *atten*; O.Norse. *eyland* - 'Place at the Island' *Eilanda* (c1150); *Neiland* (1227). The river Stour running through the village forms the county boundary between Suffolk and Essex. A loop in this river encloses a large area of land E of the village and a curved moat at the southern end forms a D-shaped island of a piece of high ground called Court Knoll, just S of the church. This was probably the earliest settlement and a likely meeting place of the Babergh Hundred Court. The area was the Norman centre of the large Domesday lordship of Eiland, which included the parishes of Wissington, Stoke, and the Horkesleys on the Essex side of the river.

NEDGING (TYE) *(4 m N of Hadleigh)* 'Neg-en'
Cosford Hundred. AS. *Hnyddinga* - 'Place of Hnydda's family'. This was later affixed by *Tye* meaning 'common pasture' *Niedinga* (DB). Nedging joined with Naughton in 1934 to become Nedging-with-Naughton.

NEEDHAM MARKET *(3 m SE of Stowmarket)* 'Need-um'
Bosmere & Claydon Hundred. AS. *nied ham* - 'Home of refuge' *Nedham* (1245); *Nedeham Markett* (1511). Although not mentioned directly in Domesday, Needham was created by the Bishop of Ely along the major road in one corner of his manor of Barking. Morley believed it to be the *Manwick* mentioned three times in Domesday - a contraction of Monkwick i.e. 'Monks' Village'. For a time merely a hamlet of nearby Barking, Needham's greater potential standing as it did by the river Gipping and on a busy main trunk road with a link to the ancient Roman road from Ipswich to Norwich, led to it enjoying subsequent expansion at Barking's expense and eventually taking its market.

NETTLESTEAD *(6 m NE of Ipswich)* 'Nettle-sted'
Bosmere & Claydon Hundred. AS. *netele stede* - 'Place where nettles grow' *Netlestedam* (DB); *Netlestede* (1215). A description which fits most parishes so it is not known why it applied so particularly here.

NEWBOURNE *(7 m SE of Ipswich)* 'New-bonn'
Carlford Hundred. AS. *nuve burn* - 'New stream' i.e. 'Stream which has changed course'. *Neubrunna* (DB). The parish takes its name from the stream (or bourn) running S from the church through the marshes to Kirton Creek. Newbourne is partly encircled by a number of ancient springs, particularly to the N. The Domesday *Haselga* (Haspley) AS. *aesp leah* - 'aspen-treed meadow' is placed in this parish E of the village.

NEWMARKET 'New Mark-ut'
Lackford Hundred. 'New market town' *novum forum* (1200); *Novum Mercatum* (1219); *La Newmarket* (1418). This place seems to have been used at first simply as a convenient meeting spot for the merchants of Exning and their customers of the surrounding district on the Icknield Way - one of the oldest routeways in England - and as such may be regarded as the New Market-place of Exning. Due to its more comercial position, Newmarket, which comprises two parishes St Mary's and All Saints, developed at the expense of its neighbour, and taking over its ancient market - though it is said a visitation of the plague to Exning hastened this transfer.

NEWTON *(3 m E of Sudbury)* 'New-t'n'
Babergh Hundred. AS. *nuve tun* - 'New farmstead' *Niwetuna* (DB). Despite their name, all three Suffolk Newtons -the one by Corton has been lost to the sea - date back to Saxon times. It

be assumed that they were initially off-shoots of nearby villages which developed in their own right and built their own church. (as with the 'stokes' and 'thorpes') There were also manors called Newton in Akenham, Creeting and Swilland villages suggesting that these never expanded in the same way to form their own identity.

NORTH COVE *(3 m E of Beccles)* 'North Cove' *and never 'Coove'*
Wangford Hundred. AS. *cofa* - 'Place of shelter' *Cove* (1204). 'North' appears a recent addition as it was called simply Cove or Cove-by-Beccles till at least the mid C14. Even Cove was not its Domesday name, for the earliest settlement was centred on the ancient earthworks at Wade or Wathe Hall. Suckling considers this to be the Domesday *Hatheburgfelda* - 'clearing by the war-fort', which, if so, takes the site back at least to Saxon times.

NORTON *(7 m ENE of Bury St. Edmunds)* 'Nor-t'n'
Blackbourn Hundred. AS. *north tun* - 'North farmstead' *Nortuna* (DB). Tostock is immediately S, with Beyton beyond, neither of which has provided evidence of former greater importance.

NOWTON *(2 m SE of Bury St. Edmunds)* 'Note-n'
Thingoe & Blackbourn Hundred. O.Norse. *naut*; AS. *tun* - 'Farm for cattle' *Newetune* (c950); *Neotuna* (DB).

OAKLEY *(3 m NNE of Eye)* 'Og-ly' *as in 'ogre*
Hartismere Hundred. AS. *ac leah* - 'Oak meadow' *Acle* (DB). The original two parishes of Great and Little Oakley were consolidated in 1449; Little Oakley has now gone along with its church.

OCCOLD *(2 m SE of Eye)* 'Ock-uld'
Hartismere Hundred. AS. *ac holt* - 'Oak copse' *Acholt* (DB). The Normans dropped the *h* as they could not pronounce it.

OFFTON *(7½ m NW of Ipswich)* 'Orf-n' *(early)*, 'Off-t'n'
Bosmere & Claydon Hundred. AS. *Offan tun* - 'Offa's farmstead' *Offetuna* (DB). S of the church at Castle Farm are the earthworks of Offton Castle on an impressive site commanding the country between the Brett and the Gipping. Its origins are believed to go back to pre-Norman times when it was known as *Pileberga* ('stockaded castle') and occupied by either King Offa of Mercia or King Offa of Essex, depending which king afterwards gave his name to the village. William de Ambli converted the old Saxon fortification into a Norman castle in 1150 during the reign of King Stephen. The parish now includes Little Bricett or Bricett Parva - 'Beorhtric's Mound' *Briticeshaga (DB)*, which was annexed to Offton in 1503 and whose church was 'down' before 1735. The 'mound' in question must surely be the moated Nunnery Mount, part of the earthworks of the Augustinian Priory, only some 900 yds. N of the parish boundary in Great Bricett, .

OLD NEWTON *(3 m NE of Stowmarket)* 'Owld Newt'n'
Stow Hundred. AS. *eald nuve tun* - 'Old New farmstead' *Neweton* (1196). Early in the C14 there were two conterminous villages of the same name or two parts of an earlier entity: then they were called *Neutone Veteris* and *Neutone Gipping*, but now the former is the Englishised Old Newton and the latter has shed its prefix and become simply Gipping. Therefore the need for the comparative 'Old' is past but remains to distinguish it from Newton in Babergh Hundred. In Saxon times it seems that Old Newton may have been known as Dagworth and Gipping as Newton: the Domesday churches here are inexplicable upon any other hypothesis. The river Gipping was probably navigable to at least Dagworth ('Daeg's property') in 600 AD.

ONEHOUSE *(1 m NW of Stowmarket)* 'One-us'
Stow Hundred. AS. *an hus* - 'Single or isolated dwelling' *Anuhus* (DB). The place-name refers to the original manor house, seat and estate of Bartholomew de Burghersh who was one of the 12 barons into whose care the Prince of Wales was committed at the Battle of Cressy in the reign of Edward III. He died at Onehouse in 1396 without male issue. The solitary situation of his residence probably gave rise to the name of the parish which, until about 300 years ago, was covered in (probably post-Domesday) wood except for a narrow strip which ascended from the valley to the Old Hall. The site is now marked by three fragments of an ancient moat. By 1270 the population had increased to the extent that a grant for a market was obtained.

ORFORD *(5 m SSW of Aldeburgh)* 'Or-fud'
Plomesgate Hundred. O.Norse. *Ore fjord* - 'Estuary by the peninsula' *Oreford* (1164). A narrow strip of land between two waters i.e. the Kings and Lantern (or Lapthorn) Marshes between the sea to the SE and the river Ore to the NW. The place is not mentioned in Domesday. Halvergate Island appears a modern spelling of Havergate, O.Norse. *Haven garta* - 'way through a river'.

OTLEY *(6 m NW of Woodbridge)* 'Ott-ly'
Carlford Hundred. AS. *Otan leah* - 'Ota's meadow' *Otelega* (DB). Several moats and earthworks here are suggestive of days when the river Lark was navigable from Martlesham Creek through Bealings, Grundisburgh, Clopton and Burgh to its head-waters at Otley.

OULTON *(2 m W of Lowestoft)* 'Olt-'n'
Mutford & Lothingland Hundred. O.Merc. *ald*; AS *tun* - Old farmstead' or 'Ala's farmstead' *Aleton* (1203). Whilst not mentioned in Domesday, its church exhibits Norman and perhaps Saxon architecture suggesting that in 1086 it formed part of Lowestoft or Flixton. It was only formed as a separate parish in the C13. Norman Scarfe suggests 'the prosperous Ala extended his estate *Aleton* by "acquiring" a major part of Flixton'.

OUSDEN *(7 m SE of Newmarket)* 'Ows-d'n'
Risbridge Hundred. AS. *Ufes denu* - 'Ufe's valley' *Uuesdana* (DB). The valley of the river Kennett heads N close to the western parish boundary.

PAKEFIELD *(Now a part of South Lowestoft)* 'Pake-field'
Mutford & Lothingland Hundred. AS. *Pacan feld* - 'Pacca's clearing' *Paggefella* (DB). Once included the now submerged Rodenhall, AS. *Roda hale* - 'Roda's shelter' *Rodenhala* (DB). A John de Rothenhale levied a fine in Pakenham during 1236.

PAKENHAM *(5 m NE of Bury St Edmunds)* 'Poik-num' *and* 'Pake-num'
Thedwastry Hundred. AS. *Pacan tun* - 'Pacca's homestead or village' *Pakenham* (962); *Pachenham* (DB).

PALGRAVE *(5 m NNW of Eye)* 'Pall-grivv'
Hartismere Hundred. AS. *pal graef* - Several suggestions as to the place-name meaning include 'Grave enclosed with palings' (Skeat) and 'Pal's (Saxon god) sacrificial trench' (Kemble), or simply 'grove where poles are obtained' (Mills) *Palegrave* (962); *Palegraua* (DB).

PARHAM *(2½ m SSE of Framlingham)* 'Pa-rum'
Plomesgate Hundred. AS. *pearran* - 'Barred-in enclosure' *Perreham* (DB). In 1734 a male skeleton, an urn, and a spearhead were said to have been found in a gravel pit here 'belonging to some Danish chieftain'. The site of the gravel pit is in a field called Fryar's Close SE of Parham

Wood, and the finds were probably Saxon. Perhaps the place-name refers to this site, or to the earthworks which surround the romantic Old Hall which is still enclosed by double moats, the inner one still wet and of great strength.

PEASENHALL *(7 m SSW of Halesworth)* 'Peez-'n-'all'
Blything Hundred. AS. *pisena healh* - 'Sheltered land where peas grow' *Pesehala* (DB). The plural Anglo-Saxon pisan for peas became pesen in medieval English and peasen in the C16.

PETTAUGH *(10 m N of Ipswich)* 'Petter' *as in 'go-getter'*
Thredling Hundred. AS. *Paotan haugh* - 'Peota's shelter or enclosure' *Petehaga* (DB). The place-name aptly describes the low-lying land near the church beside the river Deben's affluent which traverses the Roman road.

PETTISTREE *(4 m NNE of Woodbridge)* 'Pett-iss-try'
Wilford Hundred. *Poterestre* (1253). The parish was only represented by its manors at Domesday and not by an overall parish name until the C13; it also has no hall in its name. The church is dedicated to St Peter & St Paul and Morley suggests the place-name means 'Peter's Tree' - probably an old oak which stood by (or preceded) the church, under which the Black Dominican Friars preached soon after their introduction to England in 1221. Little Charsfield, *Parva Ceresfella* (DB), now a lost parish, seems to have lain across the Deben's affluent running E from Charsfield. Bing (or Byng) AS. *Byn* - 'tilled land' *Benga/Bengas* (DB) stetched from the high flat land to the S of it running down to Byng Brook, and is represented by Bing Hall. Loudham 'Lewd-um' AS. *Leod ham* - 'Leod's enclosure' *Ludham* (DB), now represented by the site of Loudham Hall W of Melton, whlist the site of Laneburh, *Laneburc* (DB), is not known.

PLAYFORD *(4 m NE of Ipswich)*
Carlford Hundred. AS. *plega forda* - 'A ford of play (or even battle)' *Plegeforda* (DB). The ford in question was probably situated where the road N through the village crosses the river Fynn just E of the ancient site at Playford Hall. The parish is (not very convincingly) suggested as an alternative site to Great Finborough for The 'Battle of Finnesburn' in 1005.

POLSTEAD *(4½ miles SW of Hadleigh)* 'Pole-stid' *as in 'pole' not 'pool'*
Babergh Hundred. AS. *pol stede* - 'Place by the pool' *Polesteda* (DB). The great expance of water opposite the church by the road junction in the village and called the village pond is undoubtably the 'pool' from which the parish gained its ancient Saxon name.

POSLINGFORD or Poslingford-with-Chipley *(2 m NE of Clare)* 'Poz-len-ford'
Risbridge Hundred. AS. *Postlinga weorth* - 'The enclosure of Postle's family' *Poslingeorda* (DB); *Poselingwrtha* (1195). One of possibly only two Suffolk parishes whose name has been altered. It appears that down to 1327 it was called Poslingworth; its change probably refers to the crossing at the southern end of the parish - where Chilton Chapel stands - of the Clare Brook by the road from Clare to Bury. A complication with this analysis is the fact that this is in the hamlet of Wentford which already represents the ford. The hamlet of Chipley (AS. *ceap leah* - 'meadow for cattle') in the N of the parish is the site of Chipley Priory.

PRESTON *(2 m NE of Lavenham)* 'Preston'
Babergh Hundred. AS. *preosta tun* - 'Farmstead of priests' *Prestetona* (DB). As priests were later everywhere, the name, to be distinctive, must have arisen when they were few. There is a Priory Farm here but this acquired its name from its acquisition by Holy Trinity Priory in Ipswich.

RAMSHOT *(5 m SE of Woodbridge)* 'Ramms-holt'
Wilford Hundred. AS. *hraemes holt* - 'Raven's (personal) or raven's copse' *Ramesholt* (DB). Probably includes Peyton, AS. *Paega tun* - 'Paega's farm' *Peituna* (DB), which Arnott locates at Peyton Hall, standing on an ancient moated site by Ramsholt Marshes.

RATTLESDEN *(5 m W of Stowmarket)* 'Rat-el-son'
Thedwestry Hundred. AS. *Hraetles denu* - 'Ratel's or Hraetel's valley' *Ratesdana; Ratlesdena* (DB). 'Hrethel, the king of the Geats and brother of Swerting, has left his name in Rattlesden in Suffolk' (Haigh). Morley also points to the fact that a marsh plant rattle-wort was called hraetelwyrt by the Saxons. The Bury monk, John Lydgate, described how stone from Normandy was shipped to England and barged up the river to Rattlesden to build Bury Abbey. The valley which carries the modest river Rat to Stowmarket to join the Gipping to Ipswich (where it becomes the Orwell) must, at that time, have painted a very different picture. It is said that the river Orwell takes its name from a spring in the parish.

RAYDON *(3½ m SSE of Hadleigh)* 'Ray-d'n'
Samford Hundred. AS. *rygen dun* - 'Hill abounding in rye' *Rienduna* (DB). Rye bread was the staple food in Suffolk during Saxon times and for long after (Edward the Confessor presented Mildenhall to Bury Abbey in order that the monks should eat wheaten rather than rye bread). The parish is comprised of two hamlets, Upper and Lower Street. This may have resulted from the fact that the Domesday *Toft* - meaning simply a 'homestead on open ground' - was placed next to Raydon, as was *Stanfelda (*Stonefield) meaning 'a stone clearing'. South Raydon is glacial-gravel and very stony.

REDE *(7 m SSW of Bury St Edmunds)* 'Reed'
Thingoe Hundred. AS. *Readan* - 'Red (Reade)'s place' *Reoda* (DB). Red's first settlement was probably the ancient moated site of Rede Hall on the northern parish boundary by a distant tributary of the Chadd Brook.

REDGRAVE *(8 m WNW of Eye)* 'Red-grave'
Hartismere Hundred. AS. *read graef* - 'Reedy pit' or AS. *hreod graef* - 'Red grave' *Redgrafe* (11c). Morley's latter definition implies some conflict possibly connected with the three Saxon cinerary urns located here.

REDISHAM (or Great Redisham) *(5 m SW of Beccles)* 'Reddshum'; 'Reddy-shum'
Wangford Hundred. AS. *Reades ham* - 'Red (Read)'s village or homestead' *Redesham* (DB). The use of the prefix 'Great' is dying out now that the parish of 'Little Redisham' has been merged with Ringsfield.

REDLINGFIELD *(3½ m SE of Eye)* 'Red-'n-ful'
Hartismere Hundred. AS. *Readelinga feld* - 'Readel's family's clearing' *Radinghefelda* (DB); *Radlingefeld* (1166). The Hoxne stream, later called Gold Brook, rises among the earthworks that enclosed the farm buildings of the Benedictine Nunnery, founded in 1120 probably on a pre-Conquest site.

RENDHAM *(4½ m E of Framlingham)* 'Renn-dum'
Plomesgate Hundred. ?AS. *Hrindan ham* - 'Hrinda's homestead' *Rimdham/Rindham* (DB).

RENDLESHAM *(5 m NE of Woodbridge)* 'Renn-dell-sum'
Loes Hundred. AS. *Rendlaes ham* - 'Rendil's homestead' *mansio Rendili* (c730). Skeat considers

Rendlesham 'very old' relative to normal Saxon place-names and almost certainly already in existence by King Redweald's reign (593-617 AD). According to Bede, the seat of the East Anglian Royal House of the Wuffingas *mansio Rendili* was at Rendlesham. Redweald, King of the East Angles, kept court here and embraced Christianity. Later (persuaded by his wife and covering his options) he is said to have had church alters both for the worship of Christ and of Woden. Sudhelm, another East Anglian king, was afterwards baptised here by Cedda, Archbishop of York. A little further S is the site of the Sutton Hoo tumuli with the world famous Saxon ship burial believed to have once contained Redweald's body. Adjoining this site is an extensive Saxon cemetery. Several sites for the Wuffingas timber 'palace' have been suggested: Field names on an 1828 map of Rendlesham include 'Great Woodenhall', 'Great Hall Wall', 'Middle Hall Wall', etc., plus Hall Walls Wood lying to the N of the parish, which, as Norman Scarfe says, seem too good to be true! The church has an extraordinarily early dedication to the C7 St. Gregory (as at Sudbury), which may again point to a location close by.

REYDON *(2 m NNW of Southwold)* 'Ra-d'n'
Blything Hundred. AS. *rygen dun* - 'Hill abounding in rye' *Rienduna* (DB). Same meaning as Raydon in Samford Hundred except that the older form *e* replaces the *a*.

RICKINGHALL INFERIOR *(8 m NE of Ixworth)* 'Rick-en-hall'
Blackbourn Hundred. AS. *Ricinga healh* - 'Sheltered corner of land of Rica's people' *Rikingahala* (DB). The place-name refers to the sheltered nook which lies in the valley on the E bank of a tributary of the river Waveney in the NE corner of the parish.

RICKINGHALL SUPERIOR *(8 m NE of Ixworth)*
Hartismere Hundred. The Hundred boundary between Blackbourn and Hartismere runs NE-SW down the middle of the southern end of Rickinghall street dividing the parishes of Inferior (NW) and Superior (SE). The northern end of the street is entirely in Superior where the boundary is provided by the stream to the NW. The above 'shelter' obviously applies only to Inferior in Blackbourn Hundred. Superior occupies the higher ground to the E which included Aldwood Green. The antiquity of this green is confirmed by the 1332 'licence to close a way leading from Merssh to Aldewodegrene'.

RINGSFIELD *(2½ m SSW of Beccles)* 'Rings-field'
Wangford Hundred. O.Norse. *Hringes*; AS. *feld* - 'Hring's clearing' *Ringesfelda* (DB). Now incorporates the parish of Little Redisham or Redisham Parva.

RINGSHALL *(4½ m S of Stowmarket)* 'Rinn-shul'
Bosmere & Claydon Hundred. O.Norse. *Hringes healh* - 'Hring's shelter' *Ringeshala* (DB). The church is thought to stand on an earlier fortified site - perhaps the Norseman Hring's sheltered home.

RISBY *(4 m NW of Bury St Edmunds)* 'Riz-bee'
Thingoe Hundred. Dan. *Hrisa (or hris) by* - 'Hrisi's (or Brushwood) farmstead' *Ringeshala* (DB).

RISHANGLES *(4 m S of Eye)* 'Rush-ang-gules' *or* 'Rish-ang-gules'
Hartismere Hundred. AS. *risc hangra* - 'Rushy hanging woods' *Risangra* (DB); *Ryshangles* (1327). It is said that rushes formerly grew here in great profusion which no doubt accounts for the strange-sounding Anglo-Saxon name meaning 'hanging wood upon the hill-side', that falls into the river Dove at Cats (?Keter's) Bridge in Thorndon.

ROUGHAM *(3½ m SE of Bury St Edmunds)* 'Ruff-um'
Thedwestry Hundred. AS. *ruh hamm* - 'Enclosure on heathland' *Ruhham* (DB); *Rougham* (1381). Probably all rough heathland in Saxon times and given to Bury Abbey by Earl Ulfketel. The manor of Eldo, AS. *Eald heal   Oldhaugh* (DB), was a grange of the Abbot.

RUMBURGH *(4 m NNW of Halesworth)* 'Rum-brer'
Blything Hundred. AS. *rum burgh* - 'Wide stronghold' or 'Strongold made of tree trunks' *Romburch* (c1050); *Ramburc* (DB). Whether 'wide' or 'made of tree trunks' (which it inevitably would have been), the place-name confidently points to a 'stronghold' here. The only possible site evident on the ground in the parish is that of the moated priory. It has been suggested that this was formerly a Roman encampment which formed a garrison to defend Stone Street which they had constructed nearby, though it is still some way to the W of this road. The Saxons would then have colonised the site, taking it for the place-name and perhaps using it for council meetings and later establishing the priory within the moat.

RUSHBROOKE *(3 m SE of Bury St Edmunds)*
Thedwestry Hundred. AS. *risc* (or *resce*) *broc* - 'Rushy' brook' *Ryscebroc* (DB). Morley considers the place-name to come from a limb of the river Lark here called Black Water which would have been broadly marshy in early times.

RUSHFORD *(4 m SE of Thetford)*
Blackbourn & Guiltcross (Norfolk) Hundreds. 'Enclosure where rushes grow' *Risseurth* (c1060); *Rusceuuorda* (DB).

RUSHMERE *(6 m ESE of Beccles)* 'Rush-mer'
Mutford & Lothingland Hundred. AS. *risc mere* - 'Rushy mere' *Ryscemara* (DB). Suckling says that 'Rushmere Hall occupied a low situation in the meadows at the south of the village' (between the Hundred River and the main road from Wrentham).

RUSHMERE ST ANDREW *(2 m NE of Ipswich)* 'Rush-mere'
Carlford Hundred. AS. *risc mere* - 'Rushy mere' plus church dedication. *Ryscemara* (DB). As William White explains 'From Bixley Decoy Ponds a rivulet flows eastward to the Deben, and has near it some poor marsh land. These ponds or meres which anciently abounded in rushes gave name to the parish.' Although Bixley (AS. *byxen leah* - 'meadow of box-trees') is now in Rushmere, Bixley Decoy is just to the S of the parish boundary in Nacton parish. However, they remain N of the Hundred boundary in Carlford whilst Nacton lies in the adjoining Hundred of Colneis.

SANTON DOWNHAM (formerly Downham) *(2 m NE of Brandon)* 'Downum'
Lackford Hundred. AS. *dun hamm* - 'Hill farmstead on sandy soil' *Dunham* (DB). Originally known as Downham, the parish later incorporated the hamlet of Santon. In 1668 a violent sandstorm blew sand from the hills of Lakenheath some five miles away burying and destroying several houses and cottages, choking the river and threatening to envelop the whole parish.

SAPISTON *(3 m NW of Ixworth)* 'Sapston' *or* 'Sapp-is-t'n'
Blackbourn Hundred. AS. *Saepes tun* - 'Sap's Farmstead' *Sapestuna* (DB). The church site (St. Andrews has a remarkable Norman S doorway) looks ripe for settlement being situated by the mill and the ford over the river Blackbourn, and in the loop of Clay Lane, an ancient track.

SAXMUNDHAM *(7 m NW of Aldeburgh)* 'Sax-mund-ham'
Plomesgate Hundred. O.Merc. *Saxmundes ham* - 'Saxmund's homestead' *Sasmundeham* (DB).

The church beside the stream probably occupies a very early Saxon situation.

SAXTEAD *(2 m NW of Framlingham)* 'Sax-tid'
Hoxne Hundred. O.Merc. *Saxan stede* - 'Seaxa's place' *Saxsteda* (DB). The parish evolved from a mere berewick (sub-manor) of Framlingham.

SEMER *(3 m NW of Hadleigh)*
Cosford Hundred. AS. *sae mere* - 'Lake or marsh-pool' *Seamera* (DB). The parish takes its name from the mere which still lies immediately S of the church. It was obviously of more considerable extent and importance in Anglo-Saxon times and the parish was known as Semere until after 1764. The mere is fed by the river Brett and this tranquil river valley is a likely spot for early settlement. Semer includes the hamlet of Ash Street - *Asce* (DB).

SHADINGFIELD *(4½ m S of Beccles)* 'Shanfeld' or 'Shadd-'n-field'
Wangford Hundred. AS. *sceathena feld* - 'Thieves' (or pirates') clearing' *Scadenafella* (DB). In 1276 it was called *Shatenefeud* meaning 'the Devil's own clearing'! The place-name probably provides evidence of the Viking marauders who were all over this district.

SHELLAND *(4 m WNW of Stowmarket)* 'Shell-und'
Stow Hundred. AS. *scelf land* - 'Land on a slope' *Sellanda* (DB). Shelland has lost its primitive *f* before the first *l* which would have originally given it the title 'Shelf-land', i.e. 'sloping-plateau' which well describes its position sloping down from the plateau of Shelland Green to Buxhall Fen on the river. Shelland church is probably that given under *Eruestuna* in 1086, and the manor of *Rockylls*, represented by Rockyll's Hall almost certainly came from the C12 tenancy of William de la Rokele, the known lord of Rockels Manor in Ringshall.

SHELLEY *(3 m SE of Hadleigh)* 'Shell-y'
Samford Hundred. AS. *scelf leah* - 'Clearing on a slope' *Sceueleia* (DB). Again, as above, this parish was originally 'shelf' and not 'shell'. The clearing contained the hall site and this upland plateau ran down to the W bank of the river Brett. The parish was once merely a berewick of Bergholt Manor.

SHIMPLING *(4½ m WNW of Lavenham)* 'Shimp-len'
Babergh Hundred. AS. *Scimplinga* - 'The settlement of Scimpel's people' *Simplinga* (DB). Otherwise unknown, this personal name *Scimpel* was possibly a nickname for 'a jester' allied to the Dutch *schimpen* 'to scoff at'. The village used to be called *Shimplingthorne* which addition of the *thorn* hamlet to the main title first appears in 1442. It is best known in the old local couplet: 'Twixt Lopham forde and Shimpling Thorn, England shalbe woonn and lorne.' Also in the parish is Chadacre, AS. *Ceddes aecer* - 'cultivated plot of a man called Cedd' *Scerdacre* (DB).

SHIPMEADOW *(3 m E of Bungay)* 'Shipp-medder'
Wangford Hundred. AS. *sceap maedwe* - 'Meadow for sheep' *Scipmedu* (DB). The old local pronounciation for sheep was 'ship' and this was often taken up in the written word. Many inland public houses bear the name 'ship' instead of 'sheep' for this reason. The old Saxon word *maed* for meadow is now sadly obsolete but is still remembered in 'the flowery mead'.

SHOTLEY *(9 m SE of Ipswich)* 'Shott-ley'
Samford Hundred. ?AS. *scota leah* - 'Clearing of huts' (Skeat) *Scoteleia* (DB). Morley has a much more interesting interpretation. He considers its meaning refers to *scot*, a Saxon fixed rent paid and noted in farms and land throughout Suffolk under the name *Scotland*. Shotley apparently paid such a tax for the making of the sea wall that keeps the Orwell within due limits. Reference

is made in the famous old ditty when Earl Bigot, fleeing the forces of Henry II to the safety of his fortress of Bungay Castle sings "I'll pay my shot to the King of Cockney". Five independant estates lay within Shotley at Domesday. Thurketelston (*Turchetlestuna* - 'Thurketel's farm'); Kirketon (*Cherchetuna* - 'Farm of the church'); Culverdeston (*Calvwetuna* - 'Ceolwulfe's farm'); Thorpe (*Torp* - 'hamlet - secondary settlement), and the curiously-named *Purte Pyt* - 'at the pit (or well)?'.

SHOTTISHAM *(4½ m SE of Woodbridge)* 'Shatsham' *or* 'Shotty-sum'
Wilford Hundred. AS. *Scottes ham* - 'Scot's homestead' *Scotesham* (DB). A building adjoining the W side of Church Lane is marked "Site of the Manor of Shottisham" on a map of 1631. The personal name Sneezum is said to be a corruption of the earlier De Shottisham.

SIBTON *(5 m NW of Saxmundham)* 'Sibb-t'n'
Blything Hundred. AS. *Sibban tun* - 'Sibba's farmstead' *Sibbetuna* (DB). The name retained its double *b* until 1275 when it became Sibeton. Sibton Hall is likely to stand on the site of the earliest settlement here when the river Yox was several feet higher than today. The lost village of Rapton (*Wrabetuna*), granted to Sibton Abbey in the C12, may have occupied the site of South Grange, one of two granges created in the parish by the Abbey, Suffolk's only Cistercian house.

SICKLESMERE *(3 m SSE of Bury St. Edmunds)*
Thedwestry Hundred. AS. *sicol mere* - 'Sickle (shaped) mere'. Describes a broadening of the river Lark on the Thingoe Hundred boundary, semi-circular in shape like a sickle, in this one-time hamlet of Great Whelnetham,

SIZEWELL *(1 m E of Leiston)* 'Size-wull'
Blything Hundred. AS. *Sisan wella* - 'Sisa's well or spring' *Syreswell* (1240). A coastal hamlet of Leiston whose Chapel of St. Nicholas was, according to Kirby, already 'down' by 1735. A well, possibly that referred to in the place-name, remains at Sizewell Hall (now in the parish of Aldringham-cum-Thorpe) right on the cliff above the North Sea.

SNAPE *(3 m SE of Saxmundham)* 'Snape'
Plomesgate Hundred. ?AS. *snaep* - 'Boggy piece of land?' or 'poor pasture?' *Snapes* (DB). Considerable doubt over the origins here especially since most of the township and the (present) hall are on heath land. *Beccinga* (DB) 'settlement of the Becclings' i.e. sons of the founder of Beccles or folk of that name, consisted of but a score of acres and is not heard of after the C13. The hamlet of Gromford, locally termed 'Grumford', AS. *grom* meaning 'fierce', tells of the days when (with our much higher river levels) our placid brooks could become brawling torrents. The crossing was at Snape Watering, NW of the church, called Thelfordon Faden's 1783 map, i.e. 'plank ford', a ford crossed by wooden planks.

SOMERLEYTON *(5 m NW of Lowestoft)* 'Somerly'; 'Summer-la-t'n' *(later)*
Mutford & Lothingland Hundred. AS. *sumerlidan tun* - 'An old pirate's farm' *Sumerledetuna* (DB). Morley suggests the place where an individual Danish pirate settled from a band who had landed in summer for plunder; later named by the Saxons 'Summer pirate's farm'. Known throughout the Middle Ages by the more attractive name of Somerley.

SOMERSHAM *(5½ m NW of Ipswich)* 'Summer-shum'
Bosmere & Claydon Hundred. AS. *sumeres hamm* - 'Enclosure for Summer' *Sumersham* (DB). Morley considers that this was an outlying tract of forest where swine-herds were sent with their stock during the warmer months but which was deserted all winter because of the cold.

SOMERTON *(7 m NE of Clare)* 'summer-t'n'
Babergh Hundred. AS. *sumer tun* - 'Summer farm' *Sumerledetuna* (DB). Similar to above.

SOTHERTON *(4 m ENE of Halesworth)* 'Sorth-et-t'n'
Blything Hundred. AS. *suthra tun* - 'Southerly farmstead' *Sudretuna* (DB). Westhall, Uggeshall and Southerton form a triangle with its base at Sotherton (the south farm).

SOTTERLEY *(5 m SSE of Beccles)* 'Sarttle-y'
Wangford Hundred. AS. *saetera leah* - 'Saetere's meadow or clearing' *Soterlega* (DB). Morley considers the place-name to be taken from the pagan Saxon god Saetere which leads him to place the unlocated Domesday vill of *Croscroft*, 'the croft beside the cross', in this parish.

SOUTH COVE *(3½ m N of Southwold)* 'South Koove'
Blything Hundred. AS. *(suth) cofa* - 'place of shelter' *Coua* (DB). Earlier the parish was termed simply *cove*, the 'South' is not much older than the 1327 Subsidy in which it first appears. Here it would no doubt refer to the (now land-locked) estuary called Easton Broad running from Long Spring in Wrentham, upon which (not the coast) its boundary abuts for a (now) uninhabited half-mile. The parish comprises two manors; *South Cove* or *North Hales* and *Polfreys* or *Blueflory Cove.*

SOUTH ELMHAM *(SW of Bungay)* 'Suth Ell-lum', *though each parish is called by its dedication alone, e.g.* S'n Krors, S'n Mar-grets, etc.
Wangford Hundred. AS. *elme hamme* - 'Enclosure near the elm tree' plus church dedications. *Almeham* (DB). However, - with the fact that South Elmham 'district' includes nine parishes each prefixed by its saint (including South Elmham St. Mary now known as Homesfield *[see Homersfield]* with a church pretty surely erected upon the site of a temple to the pagan god Hamar) forming a sub-division (ferthing) of Wangford Hundred, and with a possible Saxon 'minster' at St Cross, making this a Saxon ecclesiatical centre of great importance - Morley considers the translation 'enclosure near the elm tree' far too simplistic, though its significance cannot yet be fully understood. In neighbouring Norfolk, North Elmham again has a 'minster'. He suggests the place-name relates instead to *alm* rather than *elm* i.e. 'the gift (land) enclosure'. South Elmham was rented by Sigebert, King of the East Angles, to Felix, the Burgundian, his first bishop, who probably fixed his see at Dunwich in 630 AD. Suckling calls it a deanery within itself; but South Elmham was not a deanery; the six churches, with Sancroft, Homersfield and Flixton, were exempted from both synodals and procurations (*Victoria County History*). The diaconal court of South Elmham was independant of the archidiaconal court of Suffolk till 1540, the bishops there retained the sole right of granting probate with many equally distinctive peculiarities, all showing early ordinance. Morley suggests that its origin is most likely to have taken the form of just such a grant from an ardent convert as that, asserted above, from King Sigebeorht to our earliest Bishop, thus dated 630-4, though that grant's authenticity seems none too good. Such would be a *eleemosune* or gift of benevolence, whence the Saxons who knew Greek coined their word *AElmesse* meaning our 'alms' and 'alms-giving', abbreviated in composition (e.g. AEimes-man, an alms-man); Eleemosynary property of this kind would nataurally be termed exactly the above *aelmes ham* or without the possessive medial *-es-* Elmham - 'the gift (land) enclosure'. South Elmham St. Cross has sometimes been known as Sancroft from the family of this name who held the manor here in the C13.

SOUTHOLT *(5 m SE of Eye)* 'Suth-olt'
Hoxne Hundred. AS. *south holt* - 'South copse' *Sudholda* (DB).

SOUTHWOLD *anciently* 'Sol', *later* 'Sou-fold'
Blything Hundred. AS. *suth weald* - 'South forest' *Sudwolda* (DB). Anciently called 'Sol' (e.g. 'The Battle of Sol Bay') a name that still lingers among local fishermen. The present town stands on a treeless knoll its ancient forest, like that of Dunwich, now beneath the waves - though it appears to have existed at the Domesday Survey. Southwold is nearly an island with the North Sea on its eastern side, Buss Creek on the N, and Blyth Haven to the S.

SPEXHALL *(2 m NW of Halesworth)* 'Specks-ell'
Blything Hundred. ?AS. *specces hale* - 'Woodpecker's shelter' *Specteshale* (1197). The parish is interestingly bounded on the N by Grub Lane. Coppinger recalled that he was told that a 'bower' was excavated at both angles of its eastern junction with the Roman road around 1910 - a possible site of the place-name.

SPROUGHTON *(2 m NW of Ipswich)* 'Spraw-t'n'
Samford Hundred. AS. *sprawes tun* - 'Sprow's farmstead' *Sproeston* (DB).

STANNINGFIELD *(5 m SE of Bury St Edmunds)* 'Stann-en-field'
Thedwestry Hundred. AS. *staten feld* - 'Stony clearing'. *Stanfelda* (DB). Coldham Hall, on high ground some distance S of the church has ancient origins.

STANSFIELD *(5 m NE of Clare)* 'Stanns-field'
Risbridge Hundred. AS. *stanes feld* - 'Stan's clearing'. *Stanesfelda* (DB).

STANSTEAD *(6 m NW of Sudbury)* 'Stann-stid'
Babergh Hundred. AS. *stan stede* - 'Stony place' *Stanesteda* (DB). The place-name refers to glacial gravels which comprise the southern part of the parish. This houses the main nucleated settlement bounded on the E and W by two rivulets.

STANTON (All Saints & St John the Baptist) *(3 m NE of Ixworth)* 'Stann-t'n.
Blackbourn Hundred. AS. *stan tun* - 'Stony farmstead' *Stantuna* (DB). Probably refers to stony ground, although it may describe and relate to the remains of the Roman villa at Stanton Char.

STERNFIELD *(1½ m S of Saxmundham)* 'Starn-field'
Plomesgate Hundred. AS. *sternes feld* - 'Open ground of a man called Sterne' *Sternesfelda* (DB). The probable nickname 'Stern' adopted from the AS. adjective *styrne* - 'austere'.

STOKE ASH *(3½ m SW of Eye)* 'Stook-ash'
Hartismere Hundred. AS. *stoce aesce* - 'At the settlement by the ash' *Stoches* (DB). Here doubtless in ancient days stood some notable ashtree, conspicuous in an area encompassing Thorndon and the two Thornhams which were predominantly of brambly undergrowth rather than woodland. In the days of Saxon paganism it would have been as celebrated for pilgrimages as was later St Edmund's shrine at Bury. All the Gothic peoples venerated the Great Ash calling it *Ygg-drasill*; 'its branches', says the Prose Edda, 'extend over the whole universe and its stem bears up the earth; beneath the root, which stretches to the land of the Giants, is Mimir's well wherein all wisdom is concealed'.

STOKE BY CLARE *(5 m ESE of Haverhill)*
Risbridge Hundred. AS. *stoce Clare* - 'Outlying farmstead or hamlet (of Clare)' *Stoches* (DB).

STOKE BY NAYLAND *(2 m NE of Nayland)* 'Stoke Benn-ay-lund'
Babergh Hundred. AS. *stoc atten eyland* - 'Settlement by Nayland' *Stoc* (c970). Includes the hamlets of Thorington Street and Withermarsh Green.

STONHAM ASPALL *(6 m E of Stowmarket)* 'Stonn-um Ars-ple'
Bosmere & Claydon Hundred. AS. *stan hamm* - 'Stone enclosure' *Stanham* (DB). Disected by the Roman road to *Venta Icenorum* (A140), the original Stonham was divided into three parishes before the C14, each taking its name from from the manorial family. Until 1292 this parish was known as Stonham Lambert after its rare church dedication. Later it became Stonham Aspall from the lord of the manor Roger Aspale or Haspele. The 'stone enclosure' of the place-name may refer to the Roman bath house discovered in the parish.

STOVEN *(5 m NE of Halesworth)* 'Stuv-v'n'
Blything Hundred. Icel. and AS. *stofn* - 'A felled tree stump' *Stouone* (DB). A very early settlement site for, as Morley informs us: 'In 603 Pope Gregory ordered English churches to be erected upon the sites of former Saxon pagan temples. These Saxons were tree-worshippers which takes the place-name Stoven (*Stovin* - 'broad tree-stool whence the trunk had been cut') back to pagan days'.

STOWLANGTOFT *(6½ m ENE of Bury St Edmunds)* 'Sto Lang-tuff'
Blackbourn Hundred. AS. *stow* - 'Place of assembly or holy place' *Stou* (DB); *Stowelangetot* (C13). The name is a corruption of Stowe Langetot. The Langetot family held the manor in the C12 and C13; Richard de Langetot was lord in 1206. The Church stands within an oblong double-trenched camp with clearly defined ramparts on the eastern and northern sides of the churchyard (those to the S have been destroyed by the road). This may well have been the original 'place of assembly' which created the place-name.

STOWMARKET 'Sto' *with long 'o'. Colloqually used alone: 'I'm gooen ter Sto'.*
Stow Hundred. AS. *stow* - 'Meeting place' *Stou* (DB). 'Market' had been added to the name of Stowe by 1253, and a John "de Stowe Market" was mentioned in Patent Rolls of 1299 and a Hugo de Stowe Mercato was Vicar of Eye in 1303. The place was earlier a royal manor called 'Thorney', a name which shows that the original settlement was on an island - or more likely, peninsula - in the river Gipping. Morley suggests that it probably came from the pagan Saxon god of rain, Thunor. He it was who was said to have been 'born out of the water, to have filled the rivers and poured the milk of the cloud cows upon the thirsty earth'. The present church stands on an ancient site (perhaps once used to worship this pagan god). Thorney comprised three Domesday estates: *Ciltuna*, still represented by Chilton Hall; *Torpe* or Thorpe, a still unlocated hamlet held in 1065 by Gutumund of Haughley - suggesting it lay to the NW of the present town in the area of Tot Hill (AS. *tote hylle* - 'lookout hill') - and the now totally lost *Torstuna* or Thurston. The Hundred Court for Stow Hundred was held in the town.

STOWUPLAND *(1 m NE of Stowmarket)* 'Stup-land'
Stow Hundred. 'Stow on the plateau'. Nothing seems to quite fit this parish in Domesday Book. Although known for centuries past as *Ultim, Upland* or *Upland of Stow*, and *Stow St. Peter* (although it did not get its church until 1843), Stowupland is comprised not only of Thorney which it shares with its near neighbour Stowmarket, but other independent former Saxon estates including *Roweham* held by one *Saxo* in 1065 (which Redstone thinks is represented by Saxham Street), and *Columbers*, a manor represented by the ancient moated Columbine Hall.

STRADBROKE *(9½ m NW of Framlingham)* 'Strad-brook'
Hoxne Hundred. AS. *stede broc* - 'Place by a brook' *Statebroc* (DB). 'Strad' would be suggestive of a paved (Roman) road, but the *r* was not added until it was written as Stradbrok in the 1327 Subsidy. Battlesea Green, locally called 'Battle-sy' a (now) flat stretch of arable upland also extends westward to the Horham stream where perhaps some earlier conflict took place. The further suggestive names of Battlesea Hill and Rattlerow (?*Battlehlaw* - 'a burial mound') are near the Strad Brook.

STRADISHALL *(5½ m NW of Clare)* 'Strad-e-shul'
Risbridge Hundred. AS. *straet sele* - 'Dwelling place on the (paved) road' *Stratesella* (DB). The name 'Strade' is suggestive of a Roman road possibly on the line of the A143 which bisects the parish on a line WSW-ENE. Or it may have merely meant main road i.e. the road between the castles of Clare and Lidgate.

STRATFORD ST ANDREW *(3½ m SW of Saxmundham)* 'Stratt-fer'
Plomesgate Hundred. AS. *strate forda* - 'Ford across the main road' plus church dedication. *Straffort* (DB). As its name implies, a Roman road forded the river Alde here at a point where the present A12 crosses Stratford Bridge. With rivers much higher in those days these crossings were often fraught with difficulty.

STRATFORD ST MARY *(11 m SW of Ipswich)* 'Stratt-fudd'
Samford Hundred. AS. *strate forda* - 'Ford across the main road' plus church dedication. *Stratfort* (DB). As its name implies, and like Stratford St Andrew in NE Suffolk, a Roman road (this from Colchester) forded the river (in this case the Stour) here to enter Suffolk at a point where the same A12 crosses another Stratford Bridge.

STUSTON *(7 m SE of Ipswich)* 'Stuss-t'n'
Hartismere Hundred. AS. *Stutes tun* - 'Stut's farmstead' *Stutestuna* (DB). A nickname derived from the Saxon word *stut* 'a gnat', and probably used to describe a man's agility in battle.

STUTTON *(7½ m S of Ipswich)* 'Stutt-'in'
Samford Hundred. AS. *stut tun* - 'Village or farmstead infested with horse-flies' *Stuttuna* (DB).

SUDBOURNE *(1 m S of Aldeburgh)* 'Sudd-bunn'
Plomesgate Hundred. AS. *suth burn* - 'South stream' *Sutborne* (970); *Suburna* (DB). The manor was granted to Aethelweald, Bishop of Winchester in 970 AD. A huge parish of foreshore shingle and salt-marsh, its features much modified since the place-name was bestowed. However, what is now the extensive Sudbourne Marshes lead in from the river Alde through Chaplin's Carr, Ox Carr and Moss Carr funneling down to reach the church by a narrow stream, the probable 'south stream' more prominent in early times when river levels were much higher.

SUDBURY 'Suddbreh'
Babergh Hundred. AS. *suth burh* - 'Southern fortified settlement' *Suthbyrig* (c995); *Sutberia* (DB). Sudbury is of very ancient origins. 'It was called South-Burgh, as Norwich is said to have been called North-Burgh' wrote Kirby in 1784. Its position led to it being a place of considerable importance throughout at least the 300 years between 570 and 870 AD., and finds indicate a high status settlement as far back as the Iron Age. The Anglo-Saxon Chronicle recorded that Aelfhun, Bishop of Suffolk, died at 'Suthburgh' in 797. A section of the town's fortifications was revealed in 1992.

SUTTON *(3 m SE of Woodbridge)* 'Sutt'n'
Wilford Hundred. AS. *suth tun* - 'South farmstead' *Suthtuna* (DB). The northern point in relation to the 'south farm' would have been either Rendlesham or Ufford. During the 1984 Sutton Hoo excavations the team also unearthed Neolithic pottery and excavated Bronze Age ditches showing a long period of settlement on the site extending from Neolithic to Saxon.

SWEFLING *(3 m W of Saxmundham)* 'Swuffl'n'
Plomesgate hundred. AS. *Swaeftlinga* - 'The place of Sweftel's family' *Sueflinga* (DB); *Sueftlinges* (c1150). Not recorded elsewhere, the family name meaning 'swift' may be a nickname applied to a fast runner. Five footpaths meet at a footbridge crossing of the young river Alde north of Dernford Hall obviously the site of the ancient Dern ford.

SWILLAND *(5 m NNE of Ipswich)* 'Swill-un'
Bosmere & Claydon Hundred. AS. *swin land* - 'Land of the swine' *Suinlanda* (DB). Morley informs us that an ancient deed mentions a wood called Morsehaye here - 'the wood of a Saxon called Mor'; maybe the wood in which the swine were kept and to which the place-name refers. *Newton,* represented by Newton Hall, and probably *Bruntuna,* 'Brun (or The Brown)'s Farm' were situated in the parish.

SYLEHAM *(3½ m NW of Stradbroke)* 'Sylam' or 'Si-lum'
Hoxne Hundred. AS. *sylu hamm* - 'Enclosure by a miry place' or AS. *sae land* - 'Water Isle' *Seilam* (DB), plus *Seilanda* (DB) erroneously placed in Hartismere Hundred. Both definitions are apt here for the enclosure referred to is occupied by the church. With its round tower and Saxon long and short work, Syleham Church stands on a sandy hillock by the Waveney surrounded by marshes. A raised causeway a quarter mile long is its only connection with the road. It is believed that this is the site where, in 1174, the rebellious Earl Bigod swore submission to Henry II after threats to destroy his castle at Bungay. A similar causeway on the other side of the river in Norfolk (exactly opposite to this) runs from the river back to the main road passing through Brockdish.

TANNINGTON *(4 m NW of Framlingham)* 'Tann-er-t'n'
Hoxne Hundred. AS. *Tating tun* - 'Tat's son's farmstead?' *Tatintuna* (DB). The spelling of the parish name had been *Tat* since Anglo-Saxon times. The change to *Tann* took place sometime after 1540. Although situated in High Suffolk, usually associated with late development, the church's dedication to St Ethelbert would suggest that Tannington is a very early Saxon settlement.

TATTINGSTONE *(5 m SSW of Ipswich)* 'Tatt-es-t'n'
Samford Hundred. AS. *Tates tun* - 'Tat's (or Tata's) farmstead' *Tatistuna* (DB); *Tatingeston* (1219). If Tat's home stood on the moated site of Tattingstone Hall it now lies beneath the Alton Reservoir. A single sherd of C13-C14 pottery was found on the site before it was submerged.

THEBERTON *(2 m NE of Leiston)* 'Thebb-er-t'n'
Blything Hundred. AS. *Theodbeorhtes tun* - 'Therbert's farmstead' *Thewardetuna* (DB); *Tiberton* (1178).

THELNETHAM *(3 m NW of Botesdale)* 'Feltam' *(C18 and C19)*; 'Thel-nee-thum'
Blackbourn Hundred. ?AS. *thel Witan ham* - 'Wita's valley' or 'Wita's plank-home' *Telneteham, Telnetteham, Thelneteham and Teolftham* (DB). The various spellings in Domesday for this

parish renders the name somewhat obscure. Wita probably lived on an islet of the Waveney approachable only by a wooden plank bridge.

THORINGTON *(4 m SE of Halesworth)* 'Torr-in-t'n'
Blything Hundred. P.Norse. *Thuran*; AS. *tun* - 'Thuri's farmstead' *Toretuna* (DB). The obvious connection with the Norse god *Thor* does not apply here because the Saxons always called him Thunor.

THORNDON *(3 m SW of Eye)* 'Thorn-dun'
Hartismere Hundred. AS. *thorn dun* - 'Hill where thorn-tree grow' *Tornduna* (DB). Prickly plants once grew in profusion on the boulder-clay of the hoh upon which the church stands. West of its bridge here the river Dove splays out into a small pool, large enough in Saxon times to be termed *Stanewell* i.e. 'stony mere', giving rise to the present Standwell Green. Hestley Green, S of the village, appears a telescoped form of *hedge-seate-lea* - 'meadow by quickset Whitethorn hedge'.

THORNHAM MAGNA *(3½ m SW of Eye)* 'Thorn-um'
Hartismere Hundred. AS. *thorn ham mar* - 'Homestead where thorn-trees grow' *Thornham* (DB). Known (for no reason that is apparant) as *Thornham Pilcok* in the Nonarum of 1340, its name, along with Thornham Parva and Thorndon, suggests the district was covered with brambly undergrowth rather than wooded.

THORNHAM PARVA *(3½ m SW of Eye)* 'Thorn-um'
Hartismere Hundred. AS.*thorn ham parva* - 'Homestead where thorn-trees grow' *Thornham* (DB).

THORPE MORIEUX *(9 m SSE of Bury St Edmunds)* 'Thorp Moree' *or* 'Thorp Morex', *now* 'Thorp Mor-oo'.
Cosford Hundred. O.Fris. *thorp* - 'Outlying hamlet' or 'secondary settlement' *Thorp* (DB). It is likely that this was the outlying hamlet of the much more important (at that time) settlement at Brettenham with whom it shares the common boundary with Thedwestry Hundred. The ancient seat of the Morieux family -the manor held by Roger de Murious in 1201 - but their name was not officially suffixed to the village until after 1327. Morieux is the plural of the Old French word *moriel* meaning 'mulberry-coloured' and called 'morel' when referring to horses.

THORPENESS *(1 m N of Aldeburgh)* 'Thorpeness'
Plomesgate Hundred. Scand. *thorp* - 'Hamlet of a village (Aldringham)' *Thorp* (DB). At one time a mere hamlet of Aldringham (Aldringham-cum-thorpe), it is now an independent village.

THRANDESTON *(3 m NW of Eye)* 'Framson' *(1800s), now* 'Trans-t'n'
Hartismere Hundred. Dan. *Thrandes tun* - 'Throndr's farmstead' *Thrandeston* (c1035); *Thrandestuna, Frondestuna, Frandestuna and Strandestuna* (DB). The ancient Goswold Hall, W of the Roman road and SE of the village, was perhaps at first Godesweald i.e. 'God's wood Hall' said to have been granted by the Conqueror to Walter le Bowyer (the bowman).

THURSTON *(5 m E of Bury St Edmunds)* 'Thurs-t'n'
Thedwestry Hundred. AS. *Thures tun* - 'Thur's farmstead' *Thurstuna* (DB). Isaac Taylor ascribes this township to the Norse god Thor whom the Saxons called Thunor, making it Danish in origin.

THWAITE *(5 m SSW of Eye)* 'Twate s'n George' *(early), now* 'Twate'
Hartismere Hundred. O.Norse. *thweit* - 'A forest clearing' *Theyt* (1228). Interestingly, although the *h* is dropped in its pronounciation, it was spelt Theyt without the *w* in 1228. A Norwegian

word, showing that this was one of the earliest Viking settlements in Suffolk. Morley conjectures its development thus: 'We may date it (Thwaite) roughly at 890; the earliest recorded Norse attack on East Anglia occured in 839; and its position, so far inland and actually abutting upon the Roman road to Norwich, shows not only the completeness of their 870 conquest - or the good terms already then established with the Saxons - but also is one of our best illustrations of the broad forests obviously still clothing High Suffolk.' Some other parts of the country use this name in place of *Thorp* to signify a hamlet of a village and this may be the sole case in our county for Jocelin considered it as at one time a mere berewick of Brockford Manor. Collingsford Bridge (*Cyningesford brycg* - 'Bridge at the King's ford') which spans the river Dove on the eastern parish boundary is, however, pretty surely pure Saxon.

TIMWORTH *(3½ m N of Bury St Edmunds)* 'Timmer' *(1870) with long 'i' as in 'time',* now 'Timm-worth'
Thedwestry Hundred. AS. *Timan weorth* - 'Tima's property' *Timeworda* (DB); *Tymeworthe* (1340).

TOSTOCK *(7 m E of Bury St Edmunds)* 'Taws-tic'
Thedwestry Hundred. AS. *Tottan stoc* - 'Outlying farmstead or hamlet by the look-out place' or 'Totta's settlement' *Totestoc* (DB). The manor belonged to Brithulf, son of Leoman the Saxon, at the Conquest, and Norman Scarfe suggests that Tostock Old Hall, seat of William Berdewell in 1445, may occupy the original settlement site of the parish whose oldest spelling *Totestoc* could be interpreted 'the dairy farm with the look-out'; the church stands above where the land rises to a spur. Morley considers it a corrupted use of the older *Tottan* for a masculine personal name.

TRIMLEY ST MARTIN *(8 m ESE of Ipswich)* 'Trimm-ly'
Colneis Hundred. ?AS. *Treman leah* - 'Tryma's meadow' plus church dedication *Tremelaia* (DB). *Tryma* was probably a nickname as it means to 'confirm, strengthen or put in order' hence a man who supports and upholds. Thurstonton - *Turstanestuna* (DB) is located by Arnott 'to the E of the Trimley churches of St. Martin and St. Mary' which share the same churchyard. Trimley comprised a number of estates at Domesday. Plomeyard (*Plu(m)geard*) AS. (*?Plu*)*geard* - '(?Plu)'s yard', formed the eastern portion of Trimley Commons. Grimston (*Grimestuna*), the chief manor of St. Martin parish, was situated on the SW or Orwell slope of Trimley. Its hamlet of Thorpe (*Torp*) - Grimston-with-Thorpe - 'claimed wreck of sea' at Thorpe Stone by the Orwell. Blofield AS. blawan feld - 'windy clearing' was on the Orwell slope at the head of Walton Creek. Morston (*Morestuna*) AS. *Mor ton* - 'The farm of Mor', is represented by Morston (pronounced 'Mozt'n') Hall, an C18 house on the Orwell slope. Candlent (*Candelenta*) L. *candela lenta* 'white heat', lay upon the NE or Deben slope; its name suggesting that perhaps a small beacon stood here over the King's Fleet waters in 400 AD. Capel Hall lay by the waters of the King's Fleet. Norton (*Nortuna*) later merged with Candlent. Maistana (*Maistana*) AS. maeden stan - 'maid's stone', lay in Lower Walton. Altenston (*Alteinestuna*) constituted a separate parish until after 1420 with its own church of Alston St. John which stood 1½ miles W of Grimston Hall on the Orwell slope.

TROSTON *(6½ m NNE of Bury St Edmunds)* 'Trozt'n'
Blackbourn Hundred. Norse. *Throstinga tun* - 'Throstr's family's farm' *Trostuna* (DB). The fine tumuli here including Black Hill and Troston Mount were probably associated with these Norse settlers.

TUDDENHAM ST MARTIN *(2½ m NE of Ipswich)* 'Tudd-num'
Carlford Hundred. AS. *Tuddan ham* - 'Tudda's homestead' plus church dedication *Todenham* (DB).

TUDDENHAM ST MARY *(3 m SSE of Mildenhall)* 'Tudd-num'
Lackford Hundred. AS. *Tuddan ham* - 'Tudda's homestead' plus church dedication *Todenham* (DB). Morley states that this early Saxon township, mentioned in a charter of 854 AD, occupied a peninsula between the (then) half mile broad river Lark on the N and Herringswell Fen to the S.

TUNSTALL *(7 m NE of Woodbridge)* 'Tuns-tall'
Plomesgate Hundred. AS. *tun steall* - 'Site of farmstead' *Tunestal* (DB). Dunningworth, AS. *Dunninga wyrth* - 'property of Dunn's family' *Duniworda* (DB), now a hamlet, was once a distinct township here. A 'Dunnifers Fair' was held here making this the likely former pronounciation and suggesting there was once a *Dunnaford* 'ford of Dunn' here, probably where Snape Bridge now spans the river Alde.

UBBESTON *(6½ m SW of Halesworth)* 'Upp-st'n' *(early)*, 'Ubb-ers-t'n' *(later)*
Blything Hundred. Dan. *Ubbes tun* - 'Ubbi's farmstead' *Upbestuna* (DB); *Hubston* (1283). Morley considers it a late (c1020) Danish settlement. The church stands beside the Roman road near Ubbeston Wood on the N bank of a 40' hoh thought by many to be a Roman encampment.

UFFORD *(2½ m NNE of Woodbridge)* 'Uff-ud'
Wilford Hundred. AS. *Uffan forda* - 'Uffa's ford' *Uffeworda* (DB). The name is generally ascribed to the first of the Uffa (or Wuffa) dynasty. 'The name Wuffa, king of the East Angles, may perhaps be found at Ufford in Suffolk' (Taylor). In 1939 R. W. Maitland wrote: 'In the name of Ufford we have a definate connection with Uffa (or Wuffa); and, on the rising ground in front of Ufford Place - now parkland and nursery S and N of The Avenue (Ufford Place has gone) - considerable remains of a Saxon settlement were found about a hundred years ago, which Mr. Reginald Smith of the British Museum assigns to the later part of the sixth century. Wuffa reigned c550-586 AD'. Ufford includes the small Domesday estate of *Suggenhou* AS. *sugan ho* - 'hill-spur of the sow'. Here in 1527 stood the 'free chapel of Sogonho' ¼ mile from the Ufford's parish church.

UGGESHALL *(5½ m ENE of Halesworth)* 'Out-shull', *now* 'Uggur-shull'
Blything Hundred. Dan. *Ugges hale* - 'Uggi's corner of land' or 'Uggi's shelter' *Uggiceheala* (DB). Uggeshall Hall may stand on Uggi (a nickname for ugly)'s ancient Danish site.

WALBERSWICK *(1½ m SSW of Southwold)* 'Walders-wick' *or* 'Wallsberwick'
Blything Hundred. O.Merc. *Walhberhtes wic* - 'Walhberht's creek' *Walberdeswike* (1199). Morley suggests that the lost Domesday *Wrabbatuna* formed the earliest settlement where, according to the old ditty, "To Dunwich, Soul (Southwold) and Walberswick, we pass in at a lousy Creek". *Wrabbatuna* disappeared at the time the Blythburgh Priory Charter of 1199 and the Hundred Rolls of 1275 first spelled this township *Walberdeswyk,* a name which represent *Wealhbeorhteswic*, a Saxon name, plus *wic* 'a creek'. Here the Dunwich River formerly reached the sea.

WALDRINGFIELD *(4 m S of Woodbridge)* 'Wonnerfel' *and* 'Wunnaful'
Carlford Hundred. AS. *Waldheringa felda* - 'Waldhere's family's clearing' *Waldringfeld* (c950); *Waldingafelda* (DB). Once spelled without the *r* as the two Waldingfield parishes in Babergh Hundred. The parish was once divided, with *Minima Waldringafelda* (Lesser Waldringfield) occupying the (later deserted) heathland settlement to the N of the village.

WALPOLE *(2½ m SSW of Halesworth)* 'Wal-pole' *as in 'wallow'*
Blything Hundred. AS. *weala pol* - 'The Britons' pool' *Walepola* (DB). Morley suggests this meeting of the river Blyth with four streams running in four directions at the bottom of the valley was the *weala pol* - 'the Briton's pool', an area, then of broad marshes, where the *weala* (Britons), or foreigners (to the Saxons), had their settlement. The present church - doubtless erected upon earlier Saxon foundations - stands upon a hoh above the 'pool'.

WALSHAM LE WILLOWS *(5 m E of Ixworth)* 'Woll-sum'
Blackbourn Hundred. AS. *Weales ham* - 'Wealh's homestead' *Wal(e)sam* (DB). 'Le Willows' - in this sense meaning 'beside the willow trees' - was added much later; in fact not in use until after the Dissolution. Whereas Walpole suggests the home of a band of Britons (foreigners to the Saxons), Walsham refers to a single Briton surviving by a tributary of the Blackbourn River.

WALTON *(10 m SE of Ipswich)* 'Woll-t'n'
Colneis Hundred. AS. *weala tun* - 'Enclosure of the Romano-British' *Waletuna* (DB). *Portus Adurni*, the name of the Roman fort, suggests that this, or Colnes (from which came the name of the Hundred), may have been the Deben's name prior to Saxon times. In these early days Walton and Felixstowe were indistinguishable and the whole area, which centred on the Roman Fort, was called *Burgh*. The Saxons found the old abandoned fort provided a defendable, walled home for the Britons and named it *weala tun* which later became Walton. None of the three Domesday estates here developed manorial rights. *Oxelanda*, AS. *oxa* - 'ox', Arnott located 'near Walton Hall on the Trimley boundary'; *Gulpelea*, AS. *gylp* or *gilp* - 'boasting', whence becomes 'braggart's meadow' used as a nickname and now represented by Gulpher Hall, a large modern red-brick farmhouse by a stream running S from the Deben marshes; and *Wadgata* (Wadegate).

WANGFORD *(3 m SW of Brandon)* 'Wain-ford' *(early)*, *now* 'Wang-fud'
Lackford Hundred. AS. *waegn ford* - 'Ford that could be crossed with a wagon' *Wamforda* (DB); *Wainford* (1197). Modifications to the landscape made by the great sand-inundation of 1668 (and perhaps others) have removed any signs of a former ford here. There is little doubt, however, that at one time Wangford Fen extended much further W and across the road leading to Brandon. It was common practice to name fords according to their depth; and here used to be one just so deep that a wain (Wagon) could pass through it.

WANGFORD *(3½ m NW of Southwold)* 'Wain-ford' *(early)*, *now* 'Wang-fud'
Blything Hundred. AS. *wang ford* - 'Ford by a flat field' or 'Ford by the river Wang' *Wankeforda* (DB). The Normans wrote *nk* instead of the Saxon *ng*. *Wang* is a Saxon word for 'an open field' which suggests that the river crossing was by an open field. A possible site for the former ford is where a road leaves the A12 heading E and crosses the river to Hill Farm through the extensive Wangford Common - the 'open field' of the place name? A more likely site, however, is where four roads join the present A12 trunk road to cross the north arm of the Blyth river - which Kirby termed The Wang and which was tidal with 'salt-pans' in 1086 - in the village just SW of the church.

WANTISDEN *(7 m ENE of Woodbridge)* 'Worn-s'n' or 'Wons-den'
Plomesgate Hundred. AS. *Wantes denu* - 'Want's valley' *Wantesdena* (DB). Want's valley must refer to the the valley of a tributary of Butley River which runs E-W through the S of the parish, below the Hall and some distance S of the isolated church which has Norman origins.

WASHBROOK (formerly Belstead Magna) *(3½ m SW of Ipswich)* 'Warsh-bruk' as in 'took', or 'Washbeck'
Samford Hundred. 'Brook' means stream but so did 'Wash'! Washbrook is not mentioned in Domesday when it was called Belstead Magna - the modern Belstead formerly being Belstead Parva. *Wasebroc* first appeared in 1271 therefore it cannot have a Saxon meaning. Evidence in the form of eight tenements shown along the street line on a 1595 map in the area of Washbrook church in the hamlet of Washbrook Green, show that this was the likely original village settlement of the parish later deserted in favour of a site by the busy Roman road. 'In Washbrook was formerly another church and hamlet called *Felcherche*, AS. *felan* - 'felled church', i.e. 'church in a clearing'. 'In Felchurch lived at least four wool-assessors in 1340'. The church is said to have been sited at the crossroads at the top of the steep hill on the road to Sproughton.

WATTISFIELD *(6 m ENE of Ixworth)* 'Watchfield' *or* 'Watch-vul'
Blackbourn Hundred. AS. *Waetles feld* - 'Wastel's clearing' *Watlesfelda* (DB); *Watlesfeld* (until at least 1340); *Watesfield* (1783).

WATTISHAM *(6 m SW of Needham Market)* 'Watt-ish-um'
Cosford Hundred. AS. *Wates ham* - 'Waet's homestead' *Wecesham* (DB). The present Wattisham Hall stands on a very ancient site enclosed by a circular moat - perhaps Waet's Saxon homestead.

WENHASTON *(3 m ESE of Halesworth)* 'Wenn-az-t'n'
Blything Hundred. AS. *Wenheardes tun* - 'Wenheard's farmstead' *Wenadestuna* (DB). Morley found a C14 reference to *Wenhavestun* suggesting that the river Blyth was still navigable at this time. The hamlet of Mells, *Mealla* (DB), lay beside the river Blyth in the N of the parish. From the AS. *Mealewes*, meaning 'flour', an obvious reference to the water mill which was actually said to exist on the river here in 1065, though gone by Domesday. Mells was a considerable manor and the ruins of its Norman Chapel dedicated to St. Margaret still stands at Old Chapel Farm.

WESTERFIELD *(NE of and now partly in the borough of Ipswich)* 'Wess-er-fel'
Bosmere & Claydon Hundred. Icel. *vestri*; AS. *feld* - 'westerly clearing' *Westrefelda* (DB). Morley suggests the parish was named by the Norwegian Vikings about 840 AD., and was sited westerly of, perhaps, the Romanised Burgh.

WESTHALL *(3 m NE of Halesworth)* 'Westel', *now* 'Wess-tall'
Blything Hundred. AS. *west hale* - 'At the west shelter' *Westhala*. Not mentioned in Domesday under Westhall but it may be the unlocated *Torp* (Thorpe) comprising two manors of 50 and 20 acres. This would make it (at that time) a hamlet of Uggeshall, its name later changed to identify it from other Thorpes in the county.

WESTHORPE *(8 m N of Stowmarket)* 'Wez-thorp'
Hartismere Hundred. AS. *west thorp* - 'Westerly outlying hamlet or farmstead' *Westtorp* (DB). Morley suggests a Viking origin for this settlement of hamlet status lying west of a more major settlement which he considers was not Finningham but Stoke Ash with its celebrated 'Ash of Ritual'.

WESTLETON *(5½ m NNE of Saxmundham)* 'Wessel-t'n'
Blything Hundred. AS. *west-leodes tun* - 'Man from the west's farmstead?' *Westledestuna* (DB). Morley breaks up the Domesday word into *west*; *ledes* - 'man from the'; *tuna* - 'farmstead': 'Man from the west's farmstead'. Westleton is immediately on the old Pack Way by which the Mercians

would approach Blythburgh from the Ickneild Way in west Suffolk perhaps in 654 AD.

WESTLEY *(1 m W of Bury St Edmunds)* 'West-le, *with last syllable short*.
Thingoe Hundred. AS. *west leah* - 'West meadow or clearing' *Westlea* (DB). To the west of Bury St Edmunds, its lordship belonging to the Abbots of that town when it was known as *Westley Sextons*. They, it seems, artificially redefined its boundary which extends from the Tay Stream to the Linnet River.

WESTON *(2½ m S of Beccles)* 'Wes-t'n'
Wangford Hundred. AS. *west tun* - 'West farmstead' *Westuna* (DB). Ellough immediately east of Weston, is of Norse origin and later date so Sotterley would probably be the centre to which it was described as being to the west of.

WEST STOW *(5 m NW of Bury St Edmunds)* 'Wes-to'
Blackbourn Hundred. AS. *west stow* - 'Westerly place (of assembly) or holy place' *Stowa* (DB); *Westowe* (1254). The fact that the parish was simply *Stowa* at Domesday makes it clear that the divisions into West Stow, North Stow (now totally enveloped by Kings Forest) and (perhaps) the more easterly StowLangtoft came about later. Claude Morley suggested that the place name, which (may?) relate to a 'holy place' in the parish, points to a possible Saxon 'Temple' on the site of the Norman church.

WETHERDEN *(4 m NW of Stowmarket)* 'Wether-d'n'
Stow Hundred. AS. *weather denu* - 'Valley where wether (sheep) are kept' *Wederdena* (DB). The village and church lay in a valley carrying a stream which rises in the N of the parish and flows S and SE to join the Rat and the Gipping near Newton Bridge just N of Stowmarket.

WETHERINGSETT-CUM-BROCKFORD *(6 m SW of Eye)* 'Wether-sit'
Hartismere Hundred. AS. *Wedering saete* - 'Settlers of Weder's tribe' *Wederingaseta* (DB). No connection with 'wether' sheep as the *th* only replaced *d* in the C15. Brockford ('Broc-fud'); AS. *broc forda* - 'ford through the brook', named from the crossing of the Roman road (A140) by the river Dove (now bridged), must have been considerable in early times to warrant the place-name.

WEYBREAD *(8½ m ENE of Eye)* 'Way-bread'
Hoxne Hundred. AS. *wegbraede* - 'way-breadth' which has been translated by Skeat to 'the broad plant by the wayside' and Morley 'the broad way' in reference to the Roman road still traceable in sections from Peasenhall, through Cratfield and Fressingfield to Weybread Street. *Weibrada* (DB). The greater part of Shotford Heath now lies below the waters of a lake. The name means 'ford (over the Waveney) where one paid scot or toll'. Instead, AS. *ea stede* - 'place of water' *Isteda* (DB), formed an estate distinct from Weybread - 'Istede juxta Weybread' (1281). Copinger termed it Weybread in Weybread-Eithane (Earsham) which suggests that it lay to the SW of the parish which now forms the hamlet of Weybread Street.

WHATFIELD *(3 m NE of Hadleigh)* 'Whut-field'
Cosford Hundred. AS. *hwaete feld* - 'Open land where wheat is grown' *Watefelda* (DB). This must alway have been celebrated for the quality of its wheat as the name implies a felled part of the forest where wheat is grown. 'This town is remarkable for growing excellent Seed-Wheat' observed Kirby in 1764 and, in fact, it has often been called Wheatfield.

WHEPSTEAD *(4½ m S of Bury St Edmunds)* 'Wepp-stid'
Thingoe Hundred. AS. *Hyrpan stede* - 'Hyrpa's place' *Wepstede* (942-51); *Huepestede* (DB).

The ancient moated Manston Hall in the S of the parish represents Manston, AS. *Mannes tun* - 'Farm of Mann' *Manestuna* (DB), an estate distinct from Whepstead in 1086. It later developed manorial rights and possessed a chapel where many pilgrimages were made to 'the Maiden of Manston'.

WHERSTEAD *(1½ m S of Ipswich)* 'Wur-sted'
Samford Hundred. AS. *hwearf stede* - 'Place by a wharf or landing-place' *Weruesteda* (DB). The draft being too shallow in medieval times for them to get closer, ships anchored just short of Downham Bridge ('bridge' in this instance meaning a hard beside the river) by the river Orwell and lightered their cargo into smaller sailing hoys or barges for the final journey into Ipswich. The Saxons named the place *Weruesteda* meaning 'wharf or landing place' and this is no doubt a refence to this hard which would have been a prominent feature when they arrived. It may have earlier provided the link with the E bank for the Roman road from Stratford St Mary which passed to the S of Ipswich and crossed the river on its way to the Roman fort at Walton (by Felixstowe). Known as Redgate Hard it lies opposite evidence of the continuation of the road at Nacton. The seven estates here are still shown as farms, three of them are manors: *Painetuna* - Pannington Hall (AS. *Pagan tun*); *Torintuna* - Thorington Hall; and *Beria* - Bowen Hall.

WHITTON-CUM-THURLSTON *(NW of, and now part of Ipswich)* 'Wit-t'n' - 'Therls-t'n'
Bosmere & Claydon Hundred. AS. *Hwitinga tun* - 'Hwita's (or White's) family's farmstead' *Widituna* (DB); *Witton* (1212). Whitton was once called Whitington. Thurlston: AS. *Thurulfr* (or *Thorolfr*) *tun* - 'Thurulfr (Thorolfr)'s farmstead' *Thurulfestun* (DB).

WICKHAMBROOK *(10 m SW of Bury St Edmunds)* 'Wick-um Brook'
Risbridge Hundred. AS. *wic ham broc* - 'Homestead by a brook associated with a 'vicus' (an earlier Romano-British settlement - possibly a dairy farm)' *Wicham* (DB); *Wichambrok* (1254). 'Brook' was added after 1086 to distinguish it from the other Wickhams and refers to the young river Glem which rises here between the church and Badmondisfield Hall. The place-name is suggestive of a Roman settlement in the parish, and although finds of the period have been unearthed, no settlement as such has to date come to light. It may have associations with a road ('street' can mean Roman road) and the course of a Roman road from Cambridge to Bridge Street, Long Melford possibly passed through the parish. Although there is no obvious evidence on the ground, Wickham Street in the S of the parish, is suggestive of a likely site. The hamlet lies beside a brook, a tributary of the Glem, possible site of the earliest settlement. The parish comprises a number of greens. Close by the stream are Attleton Green ('at the town'), Nunnery, Meeting and Coltsfoot ('?*Ceoles* ford'); the outland ones are Lady's Green (near Lords Lane in Ousden), Baxters, Genesis; Ashfield and Farley (?AS. *faer leah* - 'fear or danger meadow'), Malting End, Boyden End (AS. 'Boga's valley') and Badmondisfield (now called 'Bansfield') AS. 'Beadumund's clearing' *Bademondesfelda* (DB).

WICKHAM MARKET *(5 m NNE of Woodbridge)* 'Wick-um Mar-kut'
Wilford Hundred. AS. *vic ham* - 'Homestead associated with a 'vicus (an early Romano-British settlement)' *Wikham* (DB). The 'early Romano-British settlement' to which the place-name refers is in neighbouring Hacheston. There appears to have been no early grant of a market to Wickham, but a confirmation of its existance appears in 1286; Claude Morley thinks it may have existed from Saxon times. Redstone states that Edward I granted Robert de Ufford the right to hold a market in the town before 1308, and by 1360 it had become commonly known as *Wycham*

Market. The manor of Horepole ('muddy pool') once straddled the Potsford Brook. It was later called Thorpe and represented by Thorpe Hall, the moat of which lies one mile W of the village.

WICKHAM SKEITH *(5 m SW of Eye)* 'Wick-um Skeeth'

Hartismere Hundred. AS. *wic ham* - 'Homestead associated with a 'vicus' (an earlier Romano-British settlement); Norse. *skeith* - 'race-course (or place where horses were exercised or trained)' *Wichamm* (DB); *Wicham Skeyth* (1368). The place-name 'Wickham' is suggestive of a Romano-British settlement within the vicinity. A site producing artifacts of the period has been identified in the area of Wickham Hills, but the association may simply apply to its close proximity to the A140 Ipswich to Norwich Roman road. The Scandinavian place-name *Skeith* suggests a place where horses were raced or exercised. The village green with its pond or mere and anciently known as the *Grimmer* (a name of possible Danish origin) is one likely site; another is the long meadow by the church and hall.

WILBY *(6 m NNW of Framlingham)* 'Will-beh'

Hoxne Hundred. O.Norse. *Wila bor* - 'Wili's building' or AS. *wilig beag* - 'ring of willow trees' *Wilebey* (DB). Morley considers it the Norwegian *Wila or Vila bor*, the *a* being dropped and the *or* corrupted with time to become *y*. If the latter, the site could be identified with Willow Farm which lies near the church.

WILLINGHAM ST MARY *(4 m S of Beccles)* 'Will-in-gum'

Wangford Hundred. AS. *Willinga ham* - 'Willa's family's home' plus church dedication. *Wellingaham* (DB).

WILLISHAM *(7½ m NW of Ipswich)* 'Will-i-shum'

Bosmere & Claydon Hundred. 'Wiglaf's homestead' *Wilagesham* (1198).

WINGFIELD *(7 m E of Eye)* 'Wing-field'

Hoxne Hundred. AS. *Wingan feld* - 'Winga's clearing' *Wingefeld* (1035); *Wighefelda* (DB). The hamlets of Chickering (partly in Hoxne) ?AS. *cican* - 'a chicken', and Ersham, represented by Earsham Street in the N of the parish (AS. *earhs ham* - 'mean or poor home') both possessed chapels in medieval times. The former stood near The Slades (AS 'plain or open country').

WINSTON *(1½ m S of Debenham)* 'Winns-t'n'

Thredling Hundred. AS. *Wynheres tun* - 'Wynhere's farmstead' *Winestuna* (DB). Winston Hall stands next to the church beside a strong moat which encloses an oval site above the young river Deben.

WISSETT *(2 m NW of Halesworth)* 'Wiss-ut'

Blything Hundred. AS. *wih saete*. The origins are debatable. It has been suggested as meaning 'settlers by the heathen temple' or 'settlers who had previously (elsewhere) dwelt by a heathen temple', or perhaps, less evocatively, 'fold of a man called Witta' or 'willow sticks for making a fold' *Wisseta* (DB); *Witseta* (1165).

WISSINGTON (formerly Wiston) *(1½ m W of Nayland)* 'Wiston' *or* 'Wuss-t'n'

Babergh Hundred. AS. *wisc withig tun* - 'Wigswith's (or Wigswip's) farm'. Morley suggests 'Osier-meadow farm' *Wiswythetun* (c995). Once a part of the great Domesday lordship of Eiland along with Nayland, Stoke, and the Horkesleys in Essex, it was united with Nayland in 1884.

WITHERSDALE *(6½ m NE of Stradbroke)* 'Withers-dle' *as in idle*.

Hoxne Hundred. AS. *Wetheres dale* - 'Wether's valley'. Morley considered it refers to a man's nickname derived from a sheep and not taken directly from the sheep itself. *Weresdel* (DB); *Wideresdala* (1184).

WITHERSFIELD *(2 m NW of Haverhill)* 'Withers-field'
Risbridge Hundred. AS. *Wetheres feld* - 'Open land where wethers (sheep) are kept' or 'Wether's clearing'. Again Morley goes for the personal name. *Wedresfelda* (DB). The hamlet of Hanchet AS 'land settled by Henna's family' *Haningehet* (DB), is represented by Hanchet Hall and Hanchet End.

WITNESHAM *(3½ m NE of Ipswich)* 'Witt-les-um' *or* 'Wittle-shum' - *the correct form of the place-name.*
Carlford Hundred. AS. *Hwitles ham* - 'Hwitel's (or White's) homestead', son of the settler at nearby Whitton. *Wittlesham* (DB); *Witnesham* (1254). The Domesday estate of Fynford - *Finesford, Finesforda, Finlesforda* (DB), represents the ford through the river Fynn flowing into the Deben at Martlesham - more correctly the the ford named after a Saxon called *Fynn* from whom the river is named.

WIXOE *(4 m WSW of Clare)* 'Wix-er'
Risbridge Hundred. AS. *Hwittuces hoh* - 'Hwittuc's (White's) hill-spur' *Wlteskeou* (DB); *Widekeshoo* (1205). The Saxon church stands at the southern apex of this hill-spur or *hoh* overlooking the Stour valley and marks the later Saxon settlement site in this tiny parish. The parish's correct spelling is Whixoe; the *h* having been dropped only in recent times.

WOODBRIDGE 'Wood-bridge'
Loes Hundred. AS. *wudu bricge* - 'Bridge near the wood' or 'Wooden bridge' *Oddebruge* (1050); *Wudebrige* (DB). The origins of Woodbridge's place-name are a matter of debate. Some consider that the 'wooden bridge' refers to 'a wooden bridge built over a hollow between two parks near the road to Ipswich, where, in Kirby's time, there was a house called Dry Bridge'. However, it may not refer to a bridge at all, but a 'hard' for ferries by the river (Deben) near (Kyson) Wood. Morley, with his usual enterprise and imagination, surmised that it could mean 'forest ridge, crowning the acclivity above the town and visible for many miles round' or that with the ship burial at Sutton Hoo across the water it may apply to a pagan Saxon god and mean 'Woden's ridge'. Whatever its true place-name origin the town, delightfully situated on the banks of the lovely river Deben, grew up from the important Royal Saxon settlement situated to the S at Kingston (AS. *cyninges tun* - 'King's town'). The early Saxon king was termed *cyn-ing* i.e. 'child of the nation' because he was elected by the national witanagemot or assembly of the whole (*cyn*) nation. In later Saxon times our kings habitually lived at Kingston-on-Thamas and perhaps the East Anglian kings' country seat was at Kingston-in-Woodbridge.

WOOLPIT *(6 m NW of Stowmarket)* 'Wulpet' *or* 'Wool-put'
Thedwestry Hundred. AS. *wulfpyt* - 'Pit for trapping wolves' *Wlpit* (10c); *Wlfpeta* (DB). Wolves are thought to have persisted in England till long after the Conquest and probably into Tudor times. The wolves probably inhabited the ancient Woolpit Wood, still quite extensive today.

WOOLVERSTONE *(3½ m SE of Ipswich)* 'Wool-ver-sun' *or* 'Wool-ves-t'n'
Samford Hundred. AS. *Wulfheres tun* - 'Wulfhere's farmstead' *Uluerestuna, Vluerestuna, Hulferestuna* (DB). The present Woolverstone Hall which overlooks the lovely river Orwell is now occupied by Ipswich School.

WORDWELL *(5 m NNW of Bury St Edmunds)* 'Woddle'
Blackbourn Hundred. AS. *wrida wella* - 'well by the saplings' *Wridewella* (DB). A well has been discovered in the churchyard.

WORLINGHAM *(1 m SE of Beccles)* 'Woll-'n-gam'
Wangford Hundred. AS. *Werlinga ham* - 'The homestead of Werel's people' *Werlingaham* (DB). The former township of *Worlingham Parva*, now a part of North Cove, included the ancient and evocative manor house of Wade Hall.

WORLINGTON *(1 m WSW of Mildenhall)* 'Woll-'n-t'n'
Lackford Hundred. AS. *Wrethelinga tun* - 'Farmstead by the winding stream' or 'Wrethel's family's farm' *Wirilintona* (DB); *Wridelingeton* (1201). The winding stream described above refers to the river Lark which does indeed take a very meandering course whilst forming the northern parish boundary with Mildenhall and West Row. The farmstead to which the place-name refers may have occupied the site of Wamil Hall which, although strictly in the adjoining parish of Mildenhall, is a site is of very ancient origin and, from its appearance, could at one time have been within the parish.

WORLINGWORTH *(6 m NW of Framlingham)* 'Warl-'n-wuth' *or* 'Warlingworth'
Hoxne Hundred. AS. *Wilheringa worth* - 'The property belonging to Wilhere's people' *Wilrincgawertha* (c1035); *Wyrlingwortha* (DB). It has been suggested that the original Saxon settlement was not by the church and hall but in the area of the Swan Inn.

WORTHAM *(6 m NW of Eye)* 'Wurth-um'
Hartismere Hundred. AS. *worth hamm* - 'Enclosed homestead' *Wrtham* (c950); *Wortham* (DB). With a broad frontage upon the river Waveney, Wortham once comprised the two parishes of *Wortham Southmore (or Everard)* i.e. the southerly moor or the part of which the Everards were lords, and *Wortham Eastgate (or Jervis)* i.e. at the east entrance or the part of which the Jarvis family were lords. The parishes were consolidated in 1769.

WRENTHAM *(5 m N of Southwold)* 'Renn-tham'
Blything Hundred. ?Fries. *Wrentan;* AS. *ham* - 'Wrenta's homestead' with *Wrenta*, meaning 'to grumble', used as a nickname (Skeat), or 'Hrani's homestead' from an Old Icelandic personal name *Hrani* (Morley). *Wretham* (DB); *Wrentham* (1228).

WYVERSTONE *(7 m N of Stowmarket)* 'Wy-ves-t'n'
Hartismere Hundred. AS. *Wiferthes tun* - 'Wigfrith's farmstead' *Wiuerthestune* (DB).

YAXLEY *(1½ m W of Eye)* 'Yax-ly'
Hartismere Hundred. AS. *geaces leah* - 'Cuckoo meadow' *Iachlesea* (DB).

YOXFORD *(4 m N of Saxmundham)* 'Yox-ford'
Blything Hundred. AS. *yokes forda* - 'Yoke's ford' *Gokesford* (DB); *Yokeford* (1203). As the river was named from this township it is not the obvious 'ford through the river Yox', but instead 'a ford wide enough for a yoke of oxen'. A bridge now carries the main road (A12) over the river Yox E of the village, a spot where no doubt an ancient ford once gave rise to the place-name.